Get Fit with Apple Watch

Using the Apple Watch for Health and Fitness

Allen G. Taylor

Apress®

Get Fit with Apple Watch: Using the Apple Watch for Health and Fitness

Copyright © 2015 by Allen G. Taylor

ISBN-13 (pbk): 978-1-4842-1282-0

ISBN-13 (electronic): 978-1-4842-1281-3

Managing Director: Welmoed Spahr
Lead Editor: Steve Weiss
Technical Reviewer: Jeff Tang
Editorial Board: Steve Anglin, Gary Cornell, Louise Corrigan, James T. DeWolf,
    Jonathan Gennick, Robert Hutchinson, Michelle Lowman, James Markham,
    Matthew Moodie, Jeffrey Pepper, Douglas Pundick, Ben Renow-Clarke,
    Gwenan Spearing, Matt Wade, Steve Weiss
Coordinating Editor: Kevin Walter
Copy Editor: Kim Wimpsett
Compositor: SPi Global
Indexer: SPi Global
Artist: SPi Global
Cover Photo: Martijn Vroom

Distributed to the book trade worldwide by Springer Science+Business Media New York, 233 Spring Street, 6th Floor, New York, NY 10013. Phone 1-800-SPRINGER, fax (201) 348-4505, e-mail orders-ny@springer-sbm.com, or visit www.springeronline.com. Apress Media, LLC is a California LLC and the sole member (owner) is Springer Science + Business Media Finance Inc (SSBM Finance Inc). SSBM Finance Inc is a Delaware corporation.

For information on translations, please e-mail rights@apress.com, or visit www.apress.com.

Apress and friends of ED books may be purchased in bulk for academic, corporate, or promotional use. eBook versions and licenses are also available for most titles. For more information, reference our Special Bulk Sales–eBook Licensing web page at www.apress.com/bulk-sales.

Any source code or other supplementary material referenced by the author in this text is available to readers at www.apress.com. For detailed information about how to locate your book's source code, go to www.apress.com/source-code/.

*This book is dedicated to all those who value health and fitness
enough to work on their own on a regular basis.*

Contents at a Glance

Contents

About the Author

Allen G. Taylor is a lifelong runner and practitioner of a healthy diet. He runs regularly with the Hardware Harriers of West Linn, Oregon.

Allen is the author of more than 30 books and speaks internationally on science and technology and their impact on society. Allen can be reached at allen.taylor@ieee.org. He blogs at allengtaylor.com and posts items of interest at moontube.wordpress.com. Allen's Twitter handle is @SQLwriter.

About the Technical Reviewer

Jeff Tang worked on enterprise and web app development for many years before reinventing himself to focus on building great iOS and Android apps. He had an Apple-featured, top-selling iOS app with millions of users and was recognized by Google as a Top Android Market Developer. He's the author of *Beginning Google Glass Development*, published by Apress in 2014. His favorite quote is The Man in the Arena. Jeff loves simplicity, solving puzzles, and AI. His LinkedIn profile is at https://www.linkedin.com/profile/view?id=1539384, and his e-mail is jeffxtang@gmail.com.

Introduction

If you are like me, you are willing to put in some time and effort to maintain good health and even increase your level of fitness. However, it's hard to maintain the discipline to take the necessary actions day in and day out, over the long haul. It's too easy to fill up the day with necessary activities such that when evening rolls around it's too late to fit in the exercise that you know you should do.

Fitness trackers can help to keep you motivated for a while, but when the novelty wears off, most of them find themselves in the back of a drawer. To be a consistent aid to maintaining fitness, a wearable device must be more than just a fitness tracker. This is where the Apple Watch really shows its value. Yes, it is a fitness tracker, but it is also much more than that. It is an elegant premium timepiece and a supremely convenient communication device. It also has some unique features that you cannot get with any other device, wearable or not.

In the first several short chapters of this book I will provide a brief introduction to the general features of the Apple Watch and how to use them. I will then move into the real subject area of the book: how to use the Apple Watch to help you to maintain good health and improve your physical fitness, day in and day out, over a period of months and years. The Apple Watch is an appropriate accessory at the opera or ballet just as much as it is at the gym or out on the trail.

The Apple Watch contains a comprehensive suite of built-in applications that perform many of the functions that people are accustomed to performing on their smartphones. The Apple Watch serves as an extension of a person's iPhone that eliminates the need to rummage around in a purse or reach into a pocket to pull it out and then unlock. In addition to the built-in apps, the Apple Watch is supported by several thousand third-party apps that run on your iPhone and are ported via either Bluetooth or Wi-Fi directly to your watch. A number of these apps deal with health and fitness, enabling watch wearers to tailor their daily, weekly, and monthly exercise programs to their own individual needs.

The Apple Watch sounds the opening bell to a new era, when wearable technology truly enters the mainstream. There may be no better way to appreciate where the future is flowing to than to experience the Apple Watch for yourself.

Part I

Apple Watch Basic Facts

Chapter 1

Common Features

The Apple Watch comes in three collections, aimed at three different audiences and with three different price points. Within these collections, you can buy the Apple Watch in 38 different configurations. Despite these differences, which I will discuss in Chapter 2, the three collections have much in common. The functional hardware, the software, and the physical dimensions are the same among all three.

Case Dimensions

People come in all sizes, and that variability extends to wrists. Wrist size does not necessarily have anything to do with gender. Apple Watch cases come in two sizes: a smaller one with a height of 38mm and width of 33.3mm and a larger one with a height of 42mm and a width of 35.9mm. People with smaller wrists will likely find the 38mm model more comfortable. It also weighs a little less than the 42mm model does, which may make a difference to marathon runners and others who move their arms a lot in the course of a day. The 42mm model will likely be preferred by people with larger wrists or those who want the largest display size available.

Digital Crown

Mechanical watches have had crowns longer than wristwatches have existed. Pocket watches, such as the one Abraham Lincoln wore, have crowns too. The crown of a mechanical wristwatch is a little wheel on the side that you turn to "wind" the watch. Winding turns the spring that powers the clockwork mechanism that causes the movement to advance the hands on the dial as time passes. This type of crown is strictly an analog device.

Of course, the Apple Watch does not have a mechanical movement and thus does not need to be wound. Nonetheless, people have come to expect watches to have crowns, and since the Apple Watch is really more of a computer than it is a watch, there needs to be some way to control its functions. One way to do that is using the touch-sensitive display, but another way is to give control functionality to a digital crown.

With the crown on a mechanical watch, you can do two things: set the hands to the desired time and wind the spring. With the digital crown on the Apple Watch, you can do three things: scroll to navigate across the display, zoom in or out, and tap to return to the previous screen.

With the Apple Watch on the left wrist (which is where most right-handed people will put it), the digital crown is on the right side of the watch from the vantage point of the wearer. Left-handers, on the other hand (pun intended), usually prefer to put their watch on the right wrist, where it is less likely to interfere with what they are doing with their dominant left hand. Apple handles this in a way that no traditional watch could. When setting up the watch, left handers can flip the display 180 degrees, making it easy for them to place the watch on their right wrist and operate the digital crown with their dominant left hand. An additional benefit of locating the watch on the nondominant wrist is that the touchscreen display can be manipulated with the dominant hand.

Retina Display, Gestures, and Force Touch

Steve Jobs introduced the Retina Display when he unveiled iPhone 4. He gave it a special name because the pixel density, and thus the resolution, of the display was greater than on previous Apple products. On the 38mm Apple Watch, the pixel density is 290 pixels per inch, and the display resolution is 272×340. The display on the 42mm model is a little bit sharper, with 302 pixels per inch and a resolution of 312×390. At these values, the sharpness of the display on both Apple Watch models is not quite up to that of an iPhone but is close.

The Apple Watch display responds to touch in much the same way that an iPhone does.

- Taps indicate selection or some other action, which depends on the active app.

- Vertical swipes scroll the current screen either up or down.

- Horizontal swipes move to either the previous or next page, assuming the app being displayed has a page-based interface.

- Left-edge swipes navigate back to where you came from.

The one difference between the touch capabilities of the Apple Watch and iPhones is that pinches are not supported on the Apple Watch.

Force Touch adds a new dimension to the touch-sensing capabilities of the Apple Watch. It detects the amount of force applied by a user's finger. When a press with significant force is applied, the currently running app may display a screen with options, such as to exit or pause the current activity.

Heart-Rate Sensor

People interested in improving their health know that exercise is important. What they may not know is how much exercise is the right amount for the fitness goals that they have set for themselves. Those who are trying to improve their health by losing weight will have different requirements from elite athletes who are trying to get the absolute maximum performance out of their bodies. One way to gauge how hard your body is working is to monitor your heart rate in terms of beats per minute. You want your heart rate to be high enough to move you toward your fitness goal, but not so high as to be a risk.

For several years athletes have been monitoring their heart rate by strapping a sensor to their chest, which transmits heart-rate readings to their smartphones via a Bluetooth connection. These straps are not very comfortable to wear and do not give immediate feedback on the fly because it's not a good idea to be looking at your phone while you are running or biking.

On the other hand, a quick glance at your Apple Watch can be done frequently without problems. The Apple Watch monitors blood flow in your wrist to determine heart rate. By glancing at your Apple Watch periodically while exercising, you can modify your pace or exertion level to match what you are trying to achieve.

Accelerometer

An *accelerometer* is an instrument that detects a change in motion. This could be from standing still to running, or it could be from changing direction while you are moving. It can be used to track movement in places where GPS location services are not available. The accelerometer in the Apple Watch is a key component of the system that records your movements. That information gets translated by software into meaningful fitness information, such as distance traveled and pace.

Gyroscope

A *gyroscope* is an instrument that detects motion but in a different way from the way the accelerometer works. It detects rotation around an axis. For example, when you lift your wrist to glance at your watch, it is the gyroscope that recognizes what you are doing and signals the processor to activate the display. There is no point in running the watch's battery down, powering the display when you are not looking at it.

Taken together, the accelerometer, the gyroscope, and the GPS functionality in your companion iPhone give you accurate information about where you are, where you have been, and where you are headed.

Ambient Light Sensor

The readability of a watch is going to be quite different on a sunny afternoon than it would be on a moonless night. In spite of that, you are going to want to read it easily in both cases and everywhere in between. To address this challenge, Apple has included an ambient light sensor in the Apple Watch. If it senses that the display has to compete with bright sunlight, it will crank up the luminosity. On the moonless night, your dark-adapted eyes don't need as much pixel power but still need some. Ordinary indoor lighting requires a moderate amount of light. Software in the watch will determine the optimal display power level based on the amount of light detected by the ambient light sensor.

Speaker and Microphone

The Apple Watch has both a speaker and a microphone, so you can talk to Siri, Dick Tracy–style, with the watch up in front of your mouth. When Siri responds, you will be able to hear what she has to say. For a quick answer to a question that occurs to you, this beats pulling out and unlocking your phone, hands down (er, I guess at least one hand is *up*—the one attached to the watch you are talking into).

Wi-Fi

Wi-Fi is supported. This means that when you are within range of a Wi-Fi hotspot, you can communicate with any of your other devices, such as your iPhone, that are also within range of that hotspot. When you are at home, within range of your wireless router, you don't have to carry your iPhone around. Your watch will link to it and enable you to make and receive calls right from your watch.

Bluetooth

The Apple Watch supports Bluetooth 4.0, which is the mechanism that it uses to communicate with its paired iPhone. It can also be used to communicate with other Bluetooth-enabled devices, assuming you have the apps designed to connect to the devices you want to communicate with.

Water Resistance

The Apple Watch is water-resistant rather than being waterproof. It's pretty hard to waterproof a device that has a speaker and microphone, which would not react well to being immersed. However, if you are out in the rain and get a little water on the watch face, that should not be a cause for concern. Of course, you should take reasonable steps to minimize your Apple Watch's exposure to water, as well as anything that might compromise the integrity of the watch case, such as dropping the watch on a hard surface.

Magnetic Battery Charger Cable

Although the Apple Watch claims to be water resistant, it is still a good idea to protect its innards as much as possible from liquid spills. That means you want to have as few ports of entry into its insides as possible. For this reason, Apple opted not to recharge its battery through a micro-USB port but rather with an inductive charger. The charging cable that comes with the watch attaches magnetically to the backside of the watch and terminates in a USB connector. You can recharge the watch by plugging the charger cable either into a powered device that has a USB port, such as a laptop, or into a USB power adapter.

USB Power Adapter

The USB power adapter is the same device you use to recharge your iPhone. It converts alternating current wall power to direct current at a voltage that the watch is designed to accept.

Quick Start Guide

The Apple Watch is a sophisticated device with a lot of capability. Getting the full value out of it will take some time and experimentation. To help you come up to speed as quickly as possible, a Quick Start Guide is included in the box your Apple Watch comes in.

Summary

Regardless of which of the 38 different configurations of the Apple Watch that you buy, you are getting a lot of capability in a compact package that is handily located on your wrist. All the functionality that is present on the top-of-the-line gold Apple Watch Edition is also present on the lowest-cost Apple Watch Sport. That being the case, what are the differences between the three collections and the different models within each collection? That is what Chapter 2 is about.

Chapter **2**

The Collections

Watches mean different things to different people. Some people consider a watch as just a convenient way to find out what time it is. Others consider it a fashion statement, and still others consider it to be a piece of jewelry that also happens to tell time. Smartwatch owners also have different expectations of what their watch will do for them. Apple was aware of the differences in what people are looking for in a smartwatch, and as a result, Apple offers three different families of the Apple Watch, each of which is called a Collection. Each Collection differs from the other Collections in the materials the watches are made of and in the bands that are offered.

All three families offer the same sizes of watch case: 42mm for larger wrists and 38mm for smaller wrists. As you would expect, the 38mm models feature fewer pixels on somewhat smaller faces. All the watches in all the Collections have the same functionality, regardless of case size or price. As you saw in Chapter 1, these features include the following:

- Digital crown
- Retina Display with Force Touch
- Heart-rate sensor
- Accelerometer
- Gyroscope
- Ambient light sensor
- Speaker
- Microphone
- Wi-Fi (802.11b/g/n), 2.4GHz

- Bluetooth 4.0

- Up to 18 hours of battery life

- Water resistance

The sensors and communication features enable a wide variety of useful functions to be performed by the many apps that continue to be written by third parties who desire access to the Apple Watch customer base. In Chapter 9, I will discuss a number of third-party apps that specifically relate to health and fitness.

The Three Collections

Apple has released the Apple Watch in three distinct families, called Collections, that differ from each other in target audience, appearance, options, and price. High quality in design as well as in form and finish are a top priority for all three Collections. However, all the models within a Collection are optimized for their particular audience.

- The Apple Watch Collection is designed for people who value elegance and appearance as well as functionality.

- The Apple Watch Sport Collection is aimed at people who will be wearing their watch during vigorous exercise, such as running or cycling. Functionality and lightweight are perhaps more important to these people.

- The Apple Watch Edition Collection is for people who want the best, most exclusive accessories and are willing to pay for them.

The watch faces in all three collections are rectangular with rounded corners. The screen resolution of the 42mm model is 390x312 pixels, and the resolution of the 38mm model is 340x272 pixels. Apple calls the display a Retina Display, which is also how iPhone displays are designated.

The Apple Watch Collection

The Apple Watch is designed for all-day wear. You could wear it on an early-morning run, later at the office, and in the evening at the opera. It contains all the health and fitness functionality of the watches in the Apple Watch Sport collection but in a package that conveys an impression of sophisticated elegance.

The Case

The case is constructed out of either highly polished stainless steel or space-black stainless steel. The steel is a refined 316L alloy that is cold forged to make it 80 percent harder than ordinary stainless steel. The casing is polished to a mirror finish. The space-black version has an additional layer of diamond-like carbon (DLC) that gives the case a distinctive look.

The Crystal

The crystal face of the Apple Watch really is a crystal, rather than some kind of glass. It is pure crystalline sapphire, which except for diamond is the hardest transparent material on Earth. The hardness provides the best scratch resistance available, short of having a pure diamond crystal watch face. However, it is also more brittle than the Ion-X glass used on the Apple Watch Sport, making it more prone to shattering if it suffers a hard blow, as it might in some intensive sports.

The Bands

Bands are made from three different materials: top-grain leathers, durable fluoroelastomer, or the same stainless steel used in the cases. Three distinct varieties of leather band are available, along with two kinds of steel band and one kind of fluoroelastomer band.

In addition to the different material available, there are also color choices. The leather loop comes in stone, light brown, bright blue, and black. These bands are handcrafted from Venezia leather in Arzignano, Italy, at a tannery with a history that spans five generations. Concealed magnets enable the wearer to wrap the band around the wrist for a precise fit.

The classic buckle, available only in black, is milled at the ECCO tannery in the Netherlands. The closure is a stainless steel buckle of traditional design. The classic buckle weighs less than any of the other bands that Apple offers. The combination of the stainless steel Apple Watch and the classic buckle actually weighs less than the Apple Watch Sport's aluminum case with any of the fluoroelastomer bands available with that watch.

The modern buckle comes in soft pink, brown, midnight blue, and black. Produced at a French tannery with a history that goes back more than 200 years, it has a sleek look and a two-piece magnetic closure, as well as an inner layer of Vectran for strength and stretch resistance.

The fluorelastomer bands are made out of the same material as the bands used by the Apple Watch Sport Collection. In the Apple Watch Collection, the fluorelastomer band is available in white and black. It is soft on the skin but strong and has a pin-and-tuck closure.

The link bracelet, crafted out of the same 316L stainless steel alloy as the Apple Watch case, is comprised of more than 100 components. Attention to detail is evident in both the design and the finish. It features a butterfly closure and gives you the ability to add or remove links for a perfect fit. Stainless steel and space-black stainless steel versions are offered.

The Milanese loop is a new interpretation of the classic design developed in Milan at the end of the 19th century. It is a woven stainless steel mesh that loops around, with a magnetic closure, thus assuring that the band fits the wearer's wrist exactly. In keeping with the band that inspired it, the Milanese loop comes in only one color, stainless steel.

The Apple Watch Sport Collection

If your primary reason for getting an Apple Watch is to help you improve your health and fitness, the Apple Watch Sport may be the one for you. Because the case is an aluminum alloy rather than stainless steel or gold, the Apple Watch Sport is somewhat lighter than most of the Apple Watch models and all the Apple Watch Edition models. Also, if you're going to be wearing your watch during intensive workouts, you may not want to subject an exclusive Apple Watch Edition to that kind of environment.

The Case

The case of the Apple Watch Sport is made from an aluminum alloy that is 60 percent stronger than traditional aluminum alloys, enhancing scratch resistance while maintaining a light weight on the wrist. The case is available in two colors, silver and space gray. As you would expect, the aluminum case is lighter than the stainless steel case of the Apple Watch and also lighter than the gold case of the Apple Watch Edition. For people running a marathon or some equally challenging endurance sport, that small difference in weight becomes more and more significant as the workout or race continues.

The Crystal

For this watch, the crystal is not actually crystal. It is Ion-X glass, which is similar to Corning's famous Gorilla Glass. It is likely not quite as scratch resistant as sapphire, but more importantly for people engaged in active or even extreme sports, it is a good deal more shatter-resistant. It will bend before it will break.

The Bands

For the Sport model, only one type of band is offered, the fluoroelastomer band that is also offered on the Apple Watch and the Apple Watch Edition. Five colors are offered with the Apple Watch Sport: white, blue, green, pink, and black. Although the different color bands are all made of the same material, they do not all weigh the same. This may matter for athletes who want to encumber their bodies with the absolute minimum weight during a competition. The black band weighs the least at 37 grams, and the white one weighs the most at 47 grams. The lightest "official" Apple Watch band, however, is not offered with the Apple Watch Sport. It is the classic buckle band offered with the Apple Watch. At 19 grams, it is about half the weight of the lightest band that comes standard with the Apple Watch Sport. If you are determined to minimize the weight of your Apple Watch, you are advised to purchase an Apple Watch Sport along with an extra classic buckle from the Apple Watch Collection.

The Apple Watch Edition Collection

The Apple Watch Edition is clearly for people who value exclusivity and elegance more than price. Edition Collection prices are such that only a limited number of people will be able to afford having a watch from this collection.

The Case

The case of the Apple Watch Edition is made from one of two different 18-karat gold alloys, one yellow gold and the other rose gold, formulated for hardness to resist scratches and other vicissitudes of everyday wear. The Apple Watch Edition is a premium product in every way, and the smooth lines and mirror polish of the gold case are outstanding examples of fine craftsmanship.

The Crystal

As is the case with the stainless steel Apple Watch, the crystal face of the Apple Watch Edition is pure crystalline sapphire, which except for diamond is the hardest transparent material on Earth. The hardness provides the best scratch resistance available, short of having a pure-diamond crystal watch face. However, it is also more brittle than the Ion-X glass used on the Apple Watch Sport, making it more prone to shattering if it suffers a hard blow, as it might in some intensive sports. I don't expect too many people to be wearing their Apple Watch Edition during intensive sports.

The Bands

For the Edition watches, bands are color coordinated with the cases. The white sport band is paired with the rose gold case, and the black sport band is paired with the yellow gold case. The black classic buckle and the midnight blue classic buckle are both paired with the yellow gold case. The rose gray modern buckle forms a nice complement to the rose gold case, and the bright red modern buckle goes well with the yellow gold case. Of course, if you want a different case/band combination, you can purchase additional bands separately.

Apple Watch Functions That Require an iPhone

The Apple Watch is a marvel of miniaturization, packing amazing functionality into a small package. However, the smallness of that package means that some functions that are routinely performed by smartphones will not fit into a watch strapped to a person's wrist. A Bluetooth wireless connection between the Apple Watch and an iPhone 5 or newer with at least iOS 8.2 makes the power of the iPhone available instantly by a quick glance at an upturned wrist.

There are, of course, a number of functions that the Apple Watch can perform all by itself. However, there are also things that you probably have become accustomed to your iPhone providing, which still require that you are carrying your iPhone when you access those functions through your watch.

For the initial release of the Apple Watch, running the watchOS 1 operating system, any function that requires Internet access will work only when your iPhone is within Bluetooth range of your Apple Watch. If you have updated your watch to watchOS 2, it will connect to the Internet directly, provided you are logged in to and within range of a Wi-Fi network. Phone calls and GPS will still require that the iPhone that is paired with your Apple Watch be nearby. This means that when you are out on a run without your iPhone, the Apple Watch will still be able to display your pulse rate, the number of steps you take, the distance traveled, and the calories burned. It will not, however, be able to show you the route you have run, unless the course you run is completely covered by Wi-Fi. Siri will be able to talk to you, but she will be able to perform any tasks for you that require Internet access only if you are within range of a Wi-Fi network.

Health and Fitness Applications of Apple Watch That Do Not Require an iPhone

While you are out on that run, monitoring your pulse rate, calories burned, and other variables, you will also be able to listen to the music stored on your Apple Watch. A set of Bluetooth headphones will convey the music from your watch to your ears with high-fidelity stereo sound. The main things you will miss while out without your iPhone and out of Wi-Fi range are messages, notifications, and alerts. You might be relieved to have some time free of those interruptions anyway.

Summary

The Apple Watch comes in three varieties, called Collections. They all have the same functionality, but for people primarily interested in fitness and health applications, the Apple Watch Sport Collection is most appropriate. To operate, an Apple Watch must be paired with an iPhone of no earlier vintage than iPhone 5. There are functions that can be performed only if the paired phone is within Bluetooth range, other functions that can be performed without the paired iPhone if you are within range of and are logged into a Wi-Fi network, and some functions that the watch can do all by itself when neither the paired iPhone nor Wi-Fi is nearby.

Chapter 3

Operating the Apple Watch

The Apple Watch is much more than just a timepiece, and the additional functionality that it has requires some learning by the wearer in order to unlock its full capability. The Apple Watch acts as a handy (wristy?) extension of the wearer's iPhone and as such delivers a lot of computing power and functionality in a convenient package. The minimalist design of the Apple Watch gives the user a limited number of ways to interact with it. Despite this, it can do a surprising number of things.

Suppose, like me, you have decided to wear an Apple Watch to enhance your health and fitness as well as tell you the time of day. Apps on the Apple Watch, from both Apple and from third-party developers, can keep you focused on getting a healthy amount of exercise every day. They can also record your workouts so that you can track trends in your performance as well as your vital signs. The first step in improving your health with Apple Watch is to pair it with your iPhone.

Pairing the Apple Watch with Your iPhone

For the Apple Watch to use the processing power and other resources of its owner's iPhone, it must be paired with the iPhone. Pairing sets up a two-way communication channel between the two devices. For the Apple Watch to deliver its full functionality, it must be within Bluetooth range of its paired iPhone. There are some things it can do when away from its paired iPhone, such as record the distance you have run or walked and the number of steps you have taken. However, it will not record your route, and its accuracy will be somewhat less than it would be if it had access to the GPS connection provided by your iPhone. After you pair the Apple Watch with your iPhone, you can enter inputs through the touchscreen, the digital crown, the button, and voice via Siri. The Apple Watch can send inputs to you via the screen, through the taptic engine, through audible tones, and also via Siri.

The Apple Watch works with the iPhone 5 and all newer models running iOS 8.2 and newer versions of that operating system. If you are still using an earlier iPhone, it is probably past time to upgrade anyway.

The pairing procedure is somewhat involved, consisting of a number of steps. Before you begin, you will need to make sure both your iPhone and your Apple Watch are ready to be paired and synced.

1. On your iPhone, go to Settings ➤ Bluetooth and make sure Bluetooth is on.

2. Make sure your iPhone is connected to either Wi-Fi or a cellular network.

3. Turn on your Apple Watch by pressing and holding the side button next to the digital crown until the Apple logo appears on the screen.

4. Make sure both your iPhone and your Apple Watch are charged up.

5. Keep the iPhone and Apple Watch close together during the pairing and syncing operation.

Now you can begin the pairing operation. Follow these steps:

1. Launch the Apple Watch app on your iPhone. This will display a screen similar to Figure 3-1.

Figure 3-1. Apple Watch pairing screen

2. On your Apple Watch, select the language you want the Apple Watch to use when communicating with you. A list can be scrolled through by rotating the digital crown or swiping the screen. Tap the desired language to make your selection.

3. Tap Start Pairing on both your Apple Watch and iPhone. A pairing animation that looks like a swirling cloud of molecules will appear on the watch face.

4. Hold your iPhone over the watch and center the
    pairing animation in the viewfinder on the iPhone
    screen. Keep the two devices aligned until you see
    a message saying "Your Apple Watch is Paired." If
    for some reason this doesn't work, you can pair the
    devices manually, following the instructions on your
    iPhone. Figure 3-2 shows your iPhone display after a
    successful pairing.

Figure 3-2. Your Apple Watch is now paired with your iPhone

5. On your iPhone, tap the Set Up as New Apple
    Watch option and follow the steps to get the Wrist
    Preference screen. Pick a wrist by tapping Left or
    Right on the iPhone. (If you are re-pairing a watch
    that had previously been paired but then de-paired,
    select Restore from Backup rather than Set Up as
    New Apple Watch.)

6. Read and agree to the terms and conditions.

7. Sign in with your Apple ID password.

8. Review settings for Usage and Diagnostics, Location
    Services, and Siri. They will be shared by the iPhone
    and Apple Watch. Figures 3-3 through 3-5 show
    these three steps in the process.

Figure 3-3. Allowing usage and diagnostic information to be sent to Apple

Figure 3-4. Enabling location information to appear on your watch

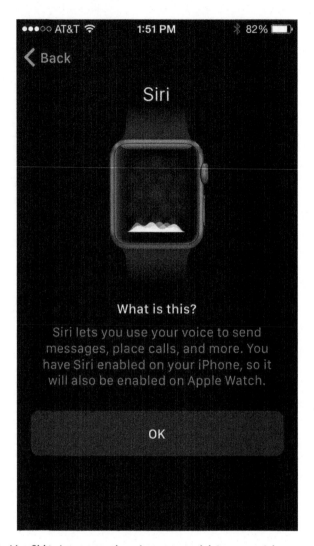

Figure 3-5. Enables Siri to hear your voice when you speak into your watch

9. On your iPhone, choose whether you want to make a passcode for your Apple Watch. You will need this to use the ApplePay feature and possibly other things. Figure 3-6 shows the Create a Passcode screen.

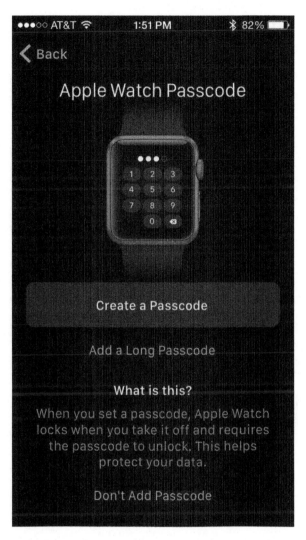

Figure 3-6. Choose whether to protect your watch with a passcode

10. As a convenience, you can elect to unlock your watch automatically whenever you unlock your iPhone. This will save some redundant typing and is still pretty safe. Figure 3-7 shows the relevant screen on your iPhone. You must approve this option on your watch.

Figure 3-7. Choose whether to unlock your watch with your iPhone

11. Sync iPhone apps that work with the Apple Watch by tapping Install All on your iPhone. If you are running a third-party fitness app, you will want to do this rather than selecting Choose Later.

12. The setup is complete. Figure 3-8 shows the screen that informs you that your Apple Watch is now ready to use.

Figure 3-8. The Apple Watch is ready

Manipulating the Touchscreen

Since it is impractical to attach either a keyboard or a mouse to a watch, the Apple Watch relies primarily on a finger interacting with the touchscreen. There are three kinds of actions you can perform with a finger: swipes, taps, and presses.

Swipes are a quick sweep of a finger across the touchscreen. They can be performed either from the top down, from the bottom up, from left to right, or from right to left. Each of these could potentially cause a different action, depending on the app. For example, if the clock face is displayed, swiping down from the top will surface your most recent notification. Swiping up will display a screen enabling you to either select or deselect Airplane Mode, Do Not Disturb Mode, or Silent Mode. It will also enable you to ping your misplaced iPhone. When you do, your phone will emit a series of ping tones, enabling you to find where you left it. Figure 3-9 shows this screen.

Figure 3-9. Handy control screen for frequently used functions

In the Activity app, with the three rings showing, swiping up from the bottom will display the calories you have burned so far today, the number of steps you have taken, and the distance you have run, walked, or cycled. Figure 3-10 shows my totals in the morning before doing much of anything. I had better get more active pronto.

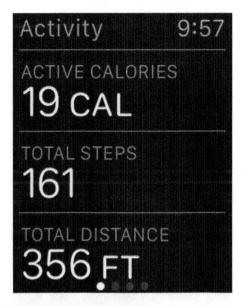

Figure 3-10. Result of swiping up in the Activity app

In the Activity app, starting at the three rings screen, swiping from right to left displays the Move ring (Figure 3-11) along with the number of calories the watch has calculated that you have burned today.

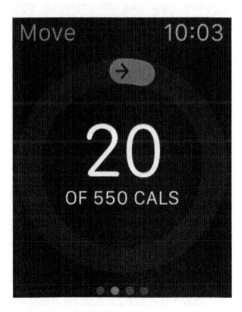

Figure 3-11. Current progress shown by the Move ring

Swiping from right to left again shows the Exercise ring (Figure 3-12), and swiping a third time shows the Stand ring (Figure 3-13).

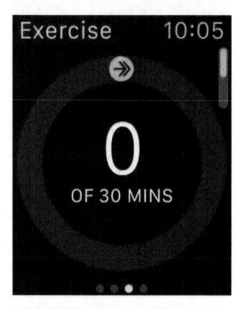

Figure 3-12. Exercise ring before today's first exercise session

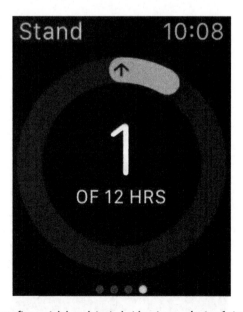

Figure 3-13. Stand ring after watch has detected at least one minute of standing so far this morning

Swiping up from any of those ring screens shows a bar graph displaying when you did your moving, exercising, or standing, on an hour-by-hour basis throughout the day.

Taps are used to make selections. On the Home screen, which displays little circular icons for the apps you have installed, tapping an icon launches the associated app. Figure 3-14 shows the Home screen on my Apple Watch, showing some of the apps I use.

Figure 3-14. Apple Watch Home screen

Force Touch, which is putting a finger on the screen and pressing, acts as an additional control; the function depends on the app that is running and the page that is currently displayed on the screen.

Twirling the Digital Crown

The digital crown, which looks a lot like the crown that you wind on a mechanical watch, can be manipulated in two different ways, either twirling it or pressing in. When you twirl the crown, what it does depends on the context. In some cases, it acts much like the wheel that sits between the buttons on a two-button mouse. Twirl it in the upward direction to cause the display to scroll up or in the downward direction to cause the display to scroll down.

Figure 3-15 shows the Home screen displayed in Figure 3-14 after it has been scrolled up.

Figure 3-15. Home screen after a scroll-up swipe

This works, of course, only if the app you are running uses up and down scrolling. You can achieve the same result by swiping the display up or down with a finger. In a different context, the crown acts as a zoom control. Twirl it up to zoom in or down to zoom out. This is the behavior you get when you twirl the crown in the Photos app. Figure 3-16 shows some recent photos in the Photo app, and Figure 3-17 shows that same screen after scrolling up.

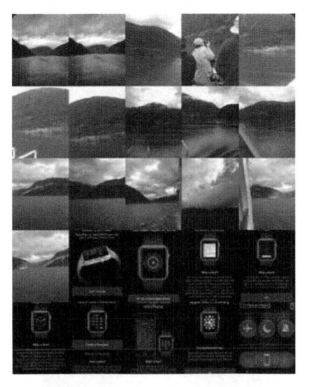

Figure 3-16. Some recent photos

Figure 3-17. Sognefjorden on a cloudy day

In the Activity app, you get the scrolling behavior when you twirl the crown rather than the zooming behavior.

Pressing the crown in takes you back to where you were before you entered the current app. For example, if you entered the Activity app by tapping the Activity icon on the clock face, pressing the crown in returns you to the clock face. If you entered the Activity app by tapping the Activity icon on the Home screen, you are returned to the Home screen.

Pressing the Button

The button located on the same side of your watch as the digital crown gives you instant access to the 12 people you call most often. Figure 3-18 shows what happens when I press this button.

Figure 3-18. Apparently I call only seven people with any regularity

By pressing the button, you can call them right from your watch, Dick Tracy–style, rather than fishing your phone out of a pocket or purse. The Friends screen displays a small icon for each of the 12 people you have identified as friends on your phone's Apple Watch app. By selecting one of them and pressing the phone handset icon, your watch will initiate a call to your friend.

If you press and hold down the button, it performs a different function, giving you the option of either powering off your watch or tapping into its power reserve. You may want to power off the watch if you don't want to be interrupted for a period of time or if you want to conserve battery power. Tapping into the power reserve gives you a little more time before the watch shuts down.

Feeling Touched by the Taptic Engine

The Apple Watch is pioneering new ground here by applying haptic feedback to a mass-market product. Haptic feedback adds another of your senses to the ways that a watch can communicate with you. Traditional mechanical watches can communicate with you in only one way—you have to look at it. Some smartwatches add a second mode of communication: sound. They emit a tone or a synthesized voice. Apple Watch adds a third, a touch that you can feel as a tap on the back of your wrist. Apple has coined the clever word *taptic* to describe the technology, combining *tap* with *haptic*.

Inside the watch, a linear actuator will discreetly tap your wrist when you receive an alert or notification. You can choose to either deal with this incoming news or ignore it for now, avoiding the interruption that pulling out your phone or even glancing at your watch would cause.

You can even communicate with other Apple Watch wearers taptically by sending them a tap or two. Since your watch is constantly monitoring your pulse, you can even send your heartbeat to someone. This expands the ways you can contact people to an entirely new dimension: voice, text, emoji, tap, and now heartbeat too.

Summary

This chapter covered how to get your Apple Watch up and running in sync with your iPhone. It also described how to control your watch with the touchscreen, as well as with the digital crown and the button. In addition to the inputs you receive from the sights and sounds you receive from the screen and the watch's speaker, the watch reaches out and touches you with its taptic engine. Apple has found ways to convey a lot of information to you through three channels in a compact package.

Chapter 4

Running the Built-in Apps

The built-in apps on the Apple Watch mirror many of the built-in apps on the iPhone but are adapted to the smaller form factor and screen size. In addition, there are new apps that are exclusive to the watch and make use of its features. In this chapter, I will give you a brief overview of the apps that correspond to iPhone apps, with special attention to differences between the two implementations. Native apps that are aimed at health and fitness, such as the Activity app and the Workout app, will be covered in detail in later chapters.

Messages

Text messaging is a two-way street. You can receive messages and also send them. With an iPhone it is possible to type a text message and send it. That is not really practical with the Apple Watch, which lacks a keyboard. However, you can record an audio message and then choose to send either the audio file or a text translation of the audio file, created by Siri's speech-to-text software.

When you receive a message on your watch, it gives you the option of replying. A large number of communications can be answered with a brief reply that falls into one of a few categories. Your watch gives you the option of replying by tapping one of a number of "canned" responses, such as those shown in Figure 4-1.

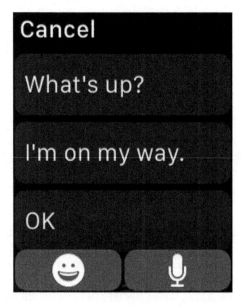

Figure 4-1. Selected responses to an incoming text message

In addition to the ones shown in Figure 4-1, the following are other possible responses:

Thank you.

Sorry. I can't talk right now.

Can I call you later?

You're welcome.

My pleasure.

No problem.

NP

Talk later?

Thanks!

An emoji

An audio message

You can select one of a number of emojis, such as the one shown in Figure 4-2.

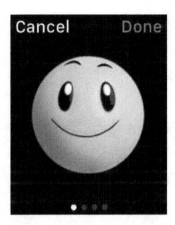

Figure 4-2. Happy face emoji

You can use the audio message to send either an audio or a text message using speech-to-text, if none of the standard replies listed previously is appropriate.

Phone

To make a call on your phone, you must pull it out of your pocket or purse and then get past your Lock screen, either with a code or a thumbprint. After selecting your phone icon, you can then choose between finding the person you want to call by specifying Favorites, Recents, Contacts, or Voicemail, or you can type in the person's phone number on your phone's keyboard.

The Apple Watch doesn't have a keyboard. However, in most cases it is a lot easier to make a call from your watch than from your phone. Your watch is right there on your wrist. Just lift up your wrist and select the Phone icon from the Home screen (Figure 4-3), and you can find the person you want to call in either Favorites, Recents, Contacts, or Voicemail (Figure 4-4) just as you can with your phone.

Figure 4-3. Phone icon on the Home screen

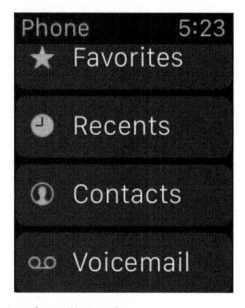

Figure 4-4. Four ways to retrieve a phone number

If the person you want to call cannot be found in any of those categories, you will just have to pull out your phone to make the call. You will very quickly start to view this as a hassle rather than as being normal. Phoning with the watch becomes natural quickly.

Mail

The small screen of the Apple Watch cannot display everything that a phone can display. When a mail message comes in to the watch, the text of the message is displayed as well as a notice to the effect that there are elements of the message that have not been displayed. Figure 4-5 shows the display when a terribly important message came in from Google Analytics.

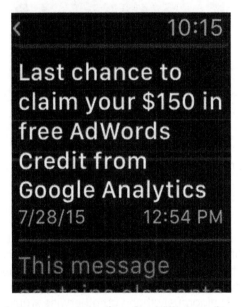

Figure 4-5. I have received a message regarding a free AdWords credit

Lacking a keyboard, the watch is not well suited for composing, typing, and sending e-mail messages. To do that, you will need to pull out your iPhone.

Calendar

The Calendar app on your Apple Watch links to the Calendar app on your iPhone. Whenever you enter an event on your phone, it will appear on your watch too. If you are using the default watch face, any event occurring today will appear there too, which is a handy reminder. You will see it whenever you glance at your watch. The Calendar app displays any events that you have already entered with your phone or other device but does not allow you to change them with the watch. Figure 4-6 shows the reminder for an upcoming appointment.

Figure 4-6. I don't want to miss this appointment with Steve

Activity

If you, like me, have bought your Apple Watch primarily as a tool to help you improve your health and fitness, then the Activity app is the one that you will use the most. In Chapter 5, I will cover the details of this health and fitness tool. Here I will only say that it is designed to get you to move your body more often and more vigorously than you otherwise would. In some ways, it acts like an addictive game, in which you constantly want to improve the best "scores" you have achieved in the past. It even supplies an assortment of "atta boys" called *achievements* to give you psychic rewards for reaching new levels of activity.

You can even brag to your friends about the latest achievements that you have earned to maintain a friendly rivalry that will improve not only your own fitness but the fitness of your Apple Watch–wearing friends as well. Come to think of it, I should text my brother right now and let him know about my latest two achievements. He probably has even more that he can tell me about and challenge me with.

Workout

The Workout app is one that you will not find on your iPhone. It uses the sensors in your Apple Watch to take data during your formal workouts. The data it takes and reports to you depend on the specific type of workout that you are doing. In Chapter 6, I will go into detail on all the different types of workouts that the app monitors. These include the following: outdoor run, outdoor walk, outdoor cycle, indoor walk, indoor run, indoor cycle, elliptical, rower, stair stepper, and the mysterious "other."

The app enables you to set goals and then records your progress toward achieving them. You can compare your performance against your best previous performance at each different activity.

Maps

The Maps app on your Apple Watch is an extension of the Maps app on your iPhone. If your phone shows a particular location and the surrounding area, your watch will show the same location but not quite as much of the surrounding area. Figure 4-7 shows a map with a pin identifying a location.

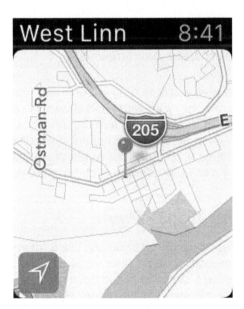

Figure 4-7. A location near Interstate 205 is pinpointed

You can use the digital crown to either zoom in or zoom out. Zooming out will give you more context, while zooming in will give you more detail. Figure 4-8 shows the pinpointed location after zooming in. It is B.J. Willy's Public House & Eatery.

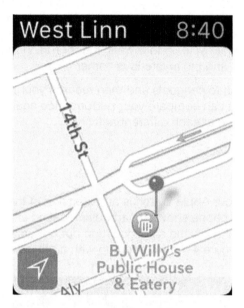

Figure 4-8. Zoomed in view of location shown in Figure 4-7

If you asked Maps to give you a recommended route from one location to another using your phone, your watch will show the same route. If you are traveling from one of those locations to the other, you can refer to your watch for directions. Even better, as the helpful voice on your phone gives you directions, the haptic engine on your Apple Watch taps your wrist three times while sounding a tone, just before each turn that you should make.

Passbook and Apple Pay

On your iPhone, Passbook is the app you can use to store electronic images of boarding passes, shopping lists, retail coupons, and merchant loyalty cards. You can also store credit and debit cards on it. This makes it convenient to board an airplane or save money at the checkout counter. It also saves you from pulling out your wallet for the appropriate piece of plastic. In fact, your wallet could become a lot thinner, with many fewer cards.

On your Apple Watch, the Passbook app syncs with the Passbook app on your phone, so any cards, passes, lists, or coupons that you have there will also appear on your watch. You can consult your shopping list while shopping. As each item is checked at the register, it is removed from the list on your watch. Figure 4-9 shows an item that is on my shopping list but has not yet been checked out at the register.

Figure 4-9. I need to buy some almond milk

To pay with Apple Pay, present your watch to the appropriate reader just like you could present your phone when making a purchase or boarding a plane. Figure 4-10 shows the card you will be paying with and the merchant you are buying from. It also serves as your loyalty card for that merchant.

Figure 4-10. Buying groceries with Apple Pay

The advantage is that a flick of the wrist and a double-click on the side button is a lot easier and faster than pulling out your phone, getting past the Lock screen, and pulling up the Passbook app. You do have to pull up the Passbook app on your Apple Watch by tapping its icon on the Home screen, but as long as your watch has been on your wrist, it assumes that you are still you and does not ask you to prove who you are.

Siri

With watchOS 2, the Siri on your watch functions the same way as Siri on your phone. As long as you are within range of a Wi-Fi signal, you can access the Internet directly from the watch. In addition, there are a few things that Siri on your watch can do that Siri on your iPhone cannot. Siri on your watch can interface directly with the watch's Activity app. Thus, you could say something like, "Hey, Siri, start a 20-minute run," and Siri would start the Workout app. You don't even need to open the app. Siri can also provide glances if you say something like "Hey, Siri, show me the Instagram glance." You can access Siri by raising your wrist to wake up the watch and then saying "Hey, Siri." Alternatively, press and hold down the digital crown until Siri asks you how she can help you. She will answer your question as text typed onto the watch face if she can, possibly accompanied by a snarky remark, or do your bidding in some other but nonetheless appropriate way. I just asked Siri what the current temperature is in Oregon City, Oregon. Figure 4-11 shows her response.

Figure 4-11. Siri delivers the current temperature in Oregon City

Wow. 104 degrees! I think I'll pass on my usual two-mile run today.

Music

One of the primary applications of the iPhone for many people is as a music player. The iTunes store offers a massive selection of music on a per-song or per-album basis. You can buy this music and play it on your phone, either through the phone's speaker or through headphones. The Music app on the Apple Watch is an extension of what you have on your iPhone. You can call up any music that you own on your phone from your watch. You don't need your phone to play music or audio, however. Your watch can store 2GB locally. The watch doesn't have a great speaker, and it doesn't have a headphone jack either. However, you can listen to high-quality audio through Bluetooth headphones, which don't need to be plugged in to work. Up-tempo music can be encouraging while you are working out at the gym or while running or cycling. Figure 4-12 shows my watch while I am listening to an up-tempo running remix. This kind of music can really encourage you to pick up your pace.

Figure 4-12. Music to run by

Camera Remote

Selfies can be a lot of fun but do not usually give you a well-composed photograph. It's hard for a photographer to hold a camera and be part of a group shot at a family gathering at the same time. A time delay works fairly well, but usually somebody in the picture will look distracted or be scratching an itch when the picture is finally taken. Camera Remote is another new app that is exclusive to the Apple Watch. You can compose your group photo, including yourself, with your iPhone set in position. The display on your watch shows you exactly what your phone camera sees. With a tap, you can either take an immediate photo or set a delay of a few seconds so that the picture does not show you tapping your watch. As long as your watch is within Bluetooth range of your phone, you can come up with all kinds of creative ways to use the Camera Remote app. Bluetooth range is about 30 feet. Figure 4-13 shows a selfie of me, not holding my phone. Since I am not tapping my Apple Watch, you can tell I opted for the three-second delay.

Figure 4-13. The hands-free selfie

Remote

Whereas the Camera Remote app enables you to control your iPhone camera from your watch, the Remote app enables you to control your Apple TV or any iTunes library that you may have on a Mac or Windows PC within Bluetooth range. I no longer need to spend time hunting for my TV remote. With the Apple Watch, my remote is now right there on my wrist. Figure 4-14 shows the first part of the instructions for setting up the Remote function.

Figure 4-14. Adding remote control capability to your watch

Weather

The iPhone gives you a full update from The Weather Channel on today's weather hour by hour, as well as current temperature, a ten-day forecast, and a number of statistics on sunrise and sunset times, chance of rain, humidity, wind, precipitation, air pressure, visibility, and UV index. The watch also gives the current temperature, ten-day forecast, weather hour by hour, sunrise and sunset and (by tapping) the chance of precipitation. Unless you are a real weather geek, the watch gives you everything you probably want to know about the weather today and for the next ten days.

That's a lot of data, but the weather statistic that concerns me the most right now is the temperature. Figure 4-15 shows the current temperature and the predicted temperatures for the next 11 hours.

Figure 4-15. *An hour before sundown, the temperature is all the way down to 102º*

Figure 4-16 shows the first few entries in the ten-day forecast.

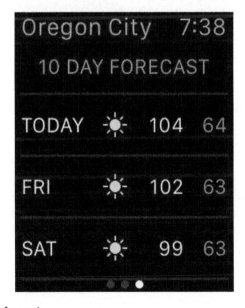

Figure 4-16. *Ten-day forecast*

Happily it looks like things are cooling off a little.

Stocks

The iPhone Stock app shows the Dow Jones, NASDAQ, and S&P 500 stock indexes, as well as the current price and daily price movement of a number of securities of your choosing. It also displays a news feed for one of those indexes or stocks, which you can change with a tap to the screen. The watch displays the same indexes and securities but does not display news feeds. There is not enough screen real estate for that. Figure 4-17 shows the major indexes. Securities are listed below them. Scroll down to see them.

Figure 4-17. The Stocks app on Apple Watch

Photos

On the iPhone, subscribers to a shared iCloud photo album can post photos and view photos posted by other subscribers. You can subscribe to multiple shared albums. The Photos app on your watch shows the most recent photos in your current album as a mosaic of tiny images. Figure 4-18 shows an example.

Figure 4-18. Lots of images crammed into a small space

By twirling the digital crown, you can zoom in on a single picture. You can also use the touchscreen to move around on the mosaic so that you can zoom in on images other than the one that happens to be located at the center of the display. Figure 4-19 shows the image one space to the right of the one in the center of Figure 4-18.

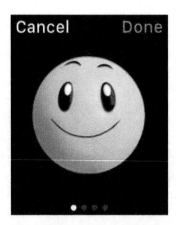

Figure 4-19. *Photos display zoomed in to a single image*

You cannot see any comments that may accompany an image or see the date it was uploaded or who has "liked" it. You may question the value of having an unordered collection of postage stamp–sized images at your beck and call, right on your wrist, but I imagine anyone separated from loved ones will appreciate being able to see the latest pictures that those loved ones have uploaded among all the other pictures in a family album. The small size of the image is probably not appropriate for fine-art photography or astrophotographs of planets and galaxies.

Alarm

On the iPhone, the Clock app has an alarm function. On the Apple Watch, the Alarm app gives you the same functionality. It enables you to set an alarm in the same way that you would set the alarm on a bedside alarm clock. Figure 4-20 shows the Edit Alarm screen that you can use to set the alarm time, the frequency of repeating, the alarm label, and whether you want to activate the Snooze function.

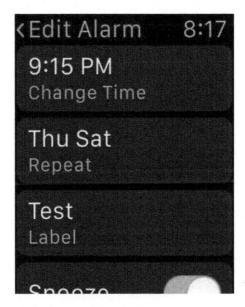

Figure 4-20. Setting an alarm on your watch

Unlike an old-fashioned alarm clock, you can name an alarm with a label and set it to repeat on a daily basis or on the specific days of the week that you choose. Like an old-fashioned alarm clock, you can also give it a snooze function if you want. You can set multiple alarms, each distinguished from the others by its label.

Stopwatch

Another function of the iPhone Clock app is the Stopwatch function. The Stopwatch app on the Apple Watch operates in pretty much the same way. You tap the green button at the lower right to start the stopwatch. It will turn red. Tap it again to stop the count. Tap the white button on the lower left to reset the display. Figure 4-21 shows the Stopwatch, with the analog display.

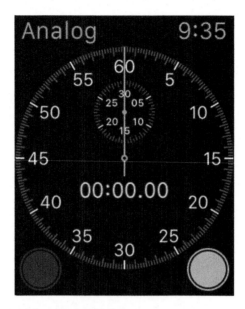

Figure 4-21. The Stopwatch is ready to start timing a run

Timer

The Timer app on the watch enables you to set a time in hours and minutes. When you tap the Start button on the screen, the display starts to count down. When the count reaches zero, a chime starts to sound and continues sounding until you dismiss it with a tap on the Dismiss button. Figure 4-22 shows the timer set to count down 15 minutes.

Figure 4-22. Countdown timer

World Clock

The World Clock function of the Clock app on the iPhone displays the current time at five cities around the world. The World Clock app on the watch gives you the current time at the same five cities, even showing whether it is daytime or nighttime at those cities. This can be helpful in telling you whether it is a good time to call in the time zone of people you might consider calling. You can add additional cities or delete existing ones on the iPhone. Your watch will reflect the changes immediately. Figure 4-23 shows that it is nighttime in Portland at 9:45 p.m. but that Portland has not been in the dark for very long. An orange dot pinpoints the location of Portland on the world map, to the right of the terminator (the curvy line separating day from night).

Figure 4-23. World Clock, showing time in Portland, Oregon

Settings

The Settings app on the Watch deals with the main things you might want to set without pulling out your phone, such as the time, Airplane Mode, Bluetooth, or the Do Not Disturb function to turn off notifications and messages. Display brightness, sound level, and haptics can also be adjusted here. You can also control whether the passcode is functional, as well as change it. Figure 4-24 shows the first several items that you can control from the Settings app.

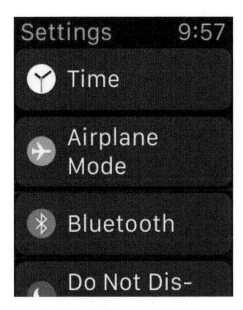

Figure 4-24. *The settings on the watch mirror those on the iPhone*

Summary

Versions of most of the built-in apps on the iPhone are also available on the Apple Watch. In addition, apps that are exclusive to the Apple Watch, such as the Exercise app, provide additional functionality. Versions of popular third-party apps are also available for download from Apple's App Store. I will discuss some of them in Chapter 9.

Part II

The Apple Watch in Health & Fitness

Chapter 5

Keeping Active with the Activity App

People have a variety of reasons for obtaining and using an Apple Watch, but the predominant reason for many is the idea that it will help them to be healthier. Scientific studies have shown that inactivity leads to ill health, while becoming more active enhances health. A device that a person wears, such as a watch, can be a powerful tool to motivate that person to be more active than they otherwise would.

What Weightlessness Does to Astronauts in Orbit

I imagine that you have heard that when astronauts come back to Earth after several months in space, they are considerably weakened. Figure 5-1 shows three astronauts who have just spent several months in orbit. They have lost bone mass and muscle and must be helped out of their space capsule and carried to the aircraft that will take them back to the space port. What causes this?

Figure 5-1. International Space Station crew 41 has landed after several months in orbit (photo courtesy of NASA)

You probably don't realize it, but for every minute of every day your body is fighting the powerful force of Earth's gravity. Ever since you were born, you have been building and maintaining strength to enable you to function in Earth's gravity field.

Astronauts in orbit do not experience that gravity field. For all intents and purposes, they are weightless. They float around within the International Space Station. They no longer have to fight the force of gravity. Their bodies adapt to the new environment by shedding bone and muscle mass that is not needed in a zero-gravity environment. This is no problem as long as they stay in space. The problem arises when they return to Earth. They must undergo months of rehabilitation before they can get back to the strength that they had before venturing into space. Although the astronauts no doubt have fun flying around the space station like Superman, they pay a price for that privilege when they return to Earth.

How Too Much Sitting Is Like Weightlessness in Space

Lying in bed or sitting in a chair is much like being weightless in space. You are not fighting gravity, at least not to the extent you would be if you were standing up, so your body adapts to the lesser load by surrendering bone mass and muscle.

Studies have shown that people with a sedentary life style in which they do not move around much are more prone to a number of degenerative diseases and tend to die at younger ages than their more active peers. The mere act of standing up and spending time in a standing position exposes you to the full force of gravity. If you spend enough time standing, your body will adapt to the need for strength by building up your bones and packing on the muscle you need to function in a one-g world.

The takeaway message here is that the more time you spend fighting the full force of gravity, the more adapted you will be to living and functioning on Earth. It's easy to fall into a habit of spending a large part of the day shielded from gravity by sitting. We sit at our desks to work. We sit on the couch to watch TV. We sit at the table to eat our meals. We sit in the stands to watch sports contests where other people are being active instead of us. To break that deadly habit, it would help to be reminded to get up and move. That's exactly what the Activity app on the Apple Watch does for us.

The Activity App: More Fun Than a Three-Ring Circus

If health and fitness were reasons for you to obtain an Apple Watch, the Activity app is the one you will be referring to most often. Not only does a quick glance tell you how far along you are toward reaching your daily activity goals, but the app also taps into everyone's natural inclination to try to excel at competitive games. The app makes a game out of choosing healthy behaviors. The display shows three concentric rings: the Move ring, the Exercise ring, and the Stand ring. Each ring tracks your performance on a different activity. Figure 5-2 shows the display after an early morning run.

Figure 5-2. The Activity app display

The Move Ring

The outermost, Move ring shows your progress toward your daily movement goal; a brightly colored dot grows into an arc that extends around the perimeter of a circle as you move. When the ring forms a complete circle, you have reached your goal. The distance traveled on the Move ring translates directly into the number of calories the watch has estimated that you have burned. The app uses the history of how you have been doing to set new weekly goals for you. If you want, you can set your own goals rather than going by what the watch thinks is best for you.

It doesn't take a math whiz to figure out that if the number of calories you consume as food and drink exceeds the number of calories you burn, you are going to gain weight. These days, the number of calories in every kind of food you can buy in the store is listed on the label or in a handy app on your phone. You can add up your daily calorie consumption to see how much you have to burn off with activity in order to maintain your current weight. Any activity beyond that should put you on the road to losing weight.

You also burn a certain number of calories just by being alive. You burn more when you move or engage in other strenuous activities. If your calorie income from food and drink exceeds your calorie outgo from basic metabolism and activity, over time you will pack on the pounds.

The purpose of the Activity app is to motivate you to make active rather than passive choices throughout the day. Take the stairs rather than the elevator. Ride your bike to the corner store rather than driving there in your car. At minor decision points throughout the day, make the more active choice. Your reward will be a payoff in terms of calories burned. As a bonus, you'll likely feel better too.

The Exercise Ring

The Exercise ring shows the number of minutes of energetic activity that you have completed. This also translates to calories burned, and the more energetic the activity, the faster the calories are burned. Any activity you engage in that causes you to exert yourself as much as you would with a brisk walk is considered exercise and causes the Exercise ring to move around the circle. What you consider to be a brisk walk may not agree with what the Apple Watch considers a brisk walk to be. I have found that walking my dog does not move the Exercise ring ahead. However, jogging with him does the trick, except for those times when he stops to examine something on the ground, which happens quite often. At least I get credit for the jogging part. The Exercise ring closes with a full circle when you have exercised for 30 minutes in a day. If you continue to exercise after achieving your goal, the ring continues to move around, gradually changing its shade of green so that you can see which loop you are on.

Any kind of exercise will do. You may decide to jog for a half hour on one day and lift weights on the next. Thirty minutes of either one will close the circle of the Exercise ring. You get the most credit for aerobic exercise. Fifteen minutes of jogging will advance the Exercise ring more than lifting weights for an hour.

The Stand Ring

The innermost, Stand ring reminds you to stand at least once for a minute or more during each of 12 hours of the day. Each time you complete one of those one-minute or more standing sessions, the Stand ring extends one-twelfth of the way around the circle. The goal is to keep your body used to the idea that you are living on a one-gravity planet rather than floating around in space. Your body needs to be constantly reminded to remain strong. If you don't remind it, the body will switch to an energy-saving mode that depletes your strength and thins your bones.

The Apple Watch can't really tell whether you are sitting or standing. It can tell only whether you have moved a significant distance in the past hour. If you stand at your standing desk for more than an hour, as I often do, the Apple Watch will remind you to stand. I solve this problem by taking a break and doing something that requires movement for at least a minute. A brief tour of the premises will do the trick nicely. The inner blue ring will advance by an hour, and you will be one-twelfth of the way closer to the daily goal of standing for at least one minute during 12 different hours of the day.

The Winner's Circle

The competitive aspect of the Activity app comes in when you look at your watch and see that it is getting late in the day and your Activity rings are nowhere near complete. You are motivated to get up out of your chair and go walk the dog, run an errand, or perform some other chore that will get you up and moving. When you come to the end of the day and all three of the rings on your Activity app are complete (Figure 5-3), you will experience a real sense of accomplishment. You will have won the game for today and in the process will have become just a little bit healthier.

Figure 5-3. Mission accomplished!

Your Unobtrusive Coach

Partway through your day, your watch will beep a couple of times, and when you glance at the screen, it will give you a report of how many calories you have burned, minutes you have exercised, and hours during which you have stood. It then encourages you to take action to move closer to your goals. Often this reminder comes at a time when you ought to take a break and do something different anyway. At the least, it gets you thinking about your plan for the rest of the day and how you might work in the activity or exercise you need to do to close all three of your rings.

Create an Activity Program Tailored to Your Needs

When it comes to maintaining health and improving fitness, everyone is starting from a unique point on the health and fitness spectrum and aspires to end up at a better point on that spectrum. What that better point is depends on what is physically possible and on the amount of effort that a person is willing to put in to achieve it. The Apple Watch can start you off on your journey to greater health by encouraging you to spend time standing, moving, and exercising. If you find that it is either consistently too easy or consistently too hard to complete one of the Activity app's three rings, you can adjust it to a level that challenges you to stretch a little but does not put completing a ring out of reach.

Ideally, standing for a least a minute in each of the 12 hours in a day will not be too difficult. It really is harmful to sit for too long. We are not designed to be sedentary. For the overwhelming majority of the existence of humanity, we have had to work practically all day at vigorous physical tasks. Sitting in an office for eight hours every day is a recent phenomenon, and our bodies are not optimized for spending so much time essentially motionless.

Standing on a regular basis is a first step (no pun intended) toward better health; moving raises the bar a notch higher, and exercising will return the most benefit for the least amount of time spent. If you complete your three rings consistently, day in and day out, you will be healthier, and you will feel better too. As a bonus, you will be more productive at whatever you do.

Chapter 6

Building Strength and Endurance with the Workout App

For the health and fitness enthusiast, this is the one to use to track and record your progress in aerobic workouts. Workout types include Outdoor Running, Outdoor Walking, Outdoor Cycle, Indoor Running, Indoor Walking, Indoor Cycle, Elliptical, Rower, Stair Stepper, and Other.

> **Note**   In Chapter 9, I will be describing third-party apps that cover some of the same ground as the built-in Workout app. In many cases, the third-party apps provide information that the built-in app does not, and vice versa. You may decide to use more than one during a workout to get a more complete picture of your performance.

Before Starting Any Workout Program

The object of working out is to improve your health and fitness, and a well-designed program will definitely do that. However, it is also possible to damage your health by engaging in a program that is not right for your body or your current state of fitness. It is always advisable to consult your doctor before making any significant change in your customary exercise program.

While you are working out, be alert to what your body is telling you. If it is telling you that it is not ready to reach the goals that you have set for yourself on your Apple Watch, back off. Gradual progress from goals that you can achieve comfortably is the best way to improve. Move from one performance plateau to the next after you have spent some time on your current plateau. Don't try to be a hero and reach too far.

How Your Progress Is Tracked

Each of the exercise types that the Workout app supports has characteristics that can be measured by the sensors in the watch. The movements that you make, how rapid and forceful they are, and your heart rate are all measured and fed into an algorithm. From those readings, plus the duration of the activity, the Workout app can estimate how many calories you have burned during the workout.

The algorithm takes into account the type of activity you are engaged in. For example, an outdoor run will burn more calories than will an outdoor walk, even if both continue for the same amount of time, say, 30 minutes. The Other category is a catchall designed to handle any activity other than the ones explicitly named. Since the Workout app does not know how vigorous your activity will be in the unknown Other activity, it makes an assumption. It assumes your calorie burn rate will be the same as it would be for a brisk walk. For this reason, the accuracy of the calorie count that it gives you will probably not be quite as good as for the known activities. Apple has done considerable research on the calorie burn rate for a variety of different types of people for each of the known workout types.

At the beginning of the workout, set your goal, or even No Goal, which will vary from one exercise type to another, and then press the virtual Start button. At that point, commence exercising. When you have reached your goal, your watch sounds a tone and taps your wrist lightly with its taptic engine. It then congratulates you on a job well done.

> **Note**   Pressing down hard on the screen will display the page with the End and Pause virtual buttons. Swiping from left to right through all the screens will also take you to the End/Pause screen. If for any reason (such as to take a phone call) you want to pause your activity, tap the Pause (double vertical bar) button. When you have finished the activity, tap the End button, which is an X icon.

Warning   Be sure to tap the X icon at the end of your workout to inform your Apple Watch that the workout has ended. If you neglect to do this, the Workout app will continue to rack up calories burned as if you were still exercising. This can lead to some ridiculous calorie burn totals that will distort your statistics. At the end of each exercise, you are given the option to either save or delete the result of the workout. You probably want to delete those for which you forgot to tap the X icon at the end of the workout. Figure 6-1 shows the workout ending screen, featuring the X that you tap to formally end the workout.

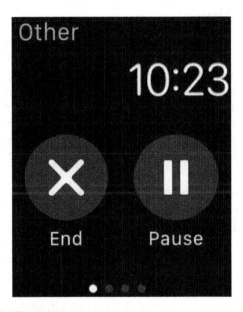

Figure 6-1. Workout End/Pause buttons

Warning   The Workout app uses a lot of electricity while it is running. This will cause your watch battery to run down, and you may not make it to the end of the day before your watch asks you if you want to switch to power reserve mode. In power reserve mode, Apple Watch shuts down all its functions except time keeping. Even the time is not displayed unless you press the button next to the digital crown. Use the Workout app judiciously, and be sure to turn it off just as soon as your workout is finished.

Supported Workouts

Different workouts make different demands on your body and burn calories at different rates. For that reason, Apple has calibrated nine of the most common activities that people do for fitness workouts. When you want to perform one of these nine workouts, you can choose it from the Workout app, and it will estimate the calories you are burning with that activity, based on the testing that Apple has previously done in its fitness lab. As you add workouts to the data available to the app, it will adjust to more closely track your actual exertion level.

Outdoor Running

Outdoor running is my favorite workout. It is highly efficient in that you can burn more calories in less time than you can with any of the other standard workouts. It has other benefits too, of course. One benefit is that it doesn't require any special equipment. If you are out in the country, the fresh country air can be a joy to breathe. The sights and sounds of the area you run through can be beautiful and relaxing. Even in the city, you can probably find a pleasant place to run. Getting away from your normal duties and concerns can free your mind to live in the moment. I have some of my best ideas while out running.

Set your goal for the exercise on your watch for the distance you want to run, the number of calories you want to burn, the amount of time you want the exercise to last, or no goal at all. Figure 6-2 shows the Outdoor Running start screen with a distance goal. You can either increase or decrease the distance you want to run by tapping the plus (+) sign or the minus (-) sign or by twirling the digital crown either up or down. This will increment or decrement the displayed distance by a fraction of a mile (or a fraction of a kilometer if your app is set to record kilometers rather than miles). You can set the distance you want to run today using repeated taps. It's a lot easier to just twirl the crown, so that's what I do. As a default, the watch will start with the most recent distance you have run and will also list the longest distance you have run previously.

Figure 6-2. Outdoor Run start screen

If the app is set, for example, to record distances in miles and for this workout you would like to use kilometers instead, just press firmly on the display. The screen will change to the MI/KM display, and you can select a new set of distance-measuring units.

If you choose to save your workouts, the Workout app will remind you of your longest run so far and thereby encourage you to exceed it. You can set your current goal to either the longest or the most recent run that you have made, or you can set it to another distance of your choice with a twirl of the digital crown.

When you start running, the app will start measuring your heart rate and will continuously monitor it during your run. You can see your current heart rate as you run by merely glancing at your watch. This is useful because you can judge your level of effort by your heart rate and adjust your speed up or down to keep it in the range that your doctor has recommended for you based on your age and fitness level. It will also tell you how long you have been running so far. Figure 6-3 shows a typical display.

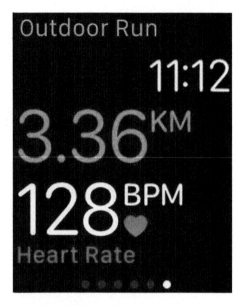

Figure 6-3. Workout screen in the middle of a run

I find that keeping tabs on my heart rate occasionally during a run keeps me in my optimal performance zone. On hilly terrain, it is interesting to see how my heart rate changes when transitioning from an uphill climb to a downhill stretch.

In addition to heart rate, the Workout app also tracks and displays elapsed time, pace, distance traveled, and calories burned. You can display whichever of those is of most interest to you on your watch while you are running. With a swipe you can move from one display to another. The app tracks and displays all this data on the fly in real time.

Outdoor Walking

If you are just getting started with an exercise program, it is a good idea to start with walking before you consider running. For the outdoor walk activity, you can set a calorie burn goal, a time of activity goal, or a distance goal as you do with the outdoor run activity. You can also choose the No Goal option if you want to record a workout but with no fixed goal in mind. Figure 6-4 shows the display for the calorie burn goal.

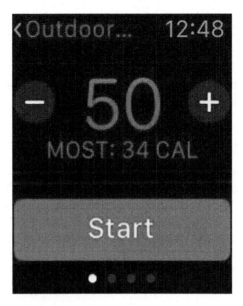

Figure 6-4. Outdoor Walk calorie burn goal

Everyone has a different level of fitness to begin, walks a different route over different terrain, and has a different preferred speed. Some people may walk with their dog, in which case the dog may have a preferred speed. Because of all these variables, it makes sense to set your calorie burn goal low at first until you can figure out what it should be for one of your typical walking sessions. As with running, don't forget to tap the End button at the end of your walk. As with the Outdoor Run app, your heart rate is continuously monitored during the exercise, and you can see it at a glance. A record is also kept that you can review later on the Health app on your iPhone.

Outdoor Cycling

The Outdoor Cycle activity is like the Outdoor Run and Outdoor Walk activities in that you can set a goal for either calories to burn, distance to travel, or time to spend on the activity, or you can leave it open ended and not set a goal. The starting display is the same as it is for outdoor running or walking and you set your goal the same way, either with the digital crown or with the plus or minus signs on the display. The information the app gives you during the exercise is the same too, including elapsed time, current speed, distance traveled so far, and heart rate.

As with the running and walking apps, pressing on the display or swiping from left to right across the display brings up the End and Pause buttons. If, for example, you ride to the store to pick up a few groceries, you can pause the workout before going into the store and then resume it when you mount your bike to head back home. This way, you can get credit for a workout and perform some useful task at the same time.

Indoor Walking

Walking is a good exercise, even if it is not convenient to do it outside, either because of bad weather or for some other reason. This is where a treadmill comes in handy. If you have one at home or a membership at a local gym, you can do your walking workout inside, where the weather is always clement. If you don't have access to a treadmill, you can achieve much of the same effect by walking around in circles. That's what I do when I want the Activity app to recognize the fact that I have stood up for at least a minute.

As I mentioned in Chapter 5, the Activity app cannot recognize whether you are sitting or standing if you have been standing still for a long time, as I do at my stand-up desk. To get the Activity app to recognize the fact that I am indeed on my feet, I walk in a loop from my kitchen to the dining room to the living room and then back to the kitchen again. When I do this several times, the app recognizes that I have indeed stood up for at least a minute.

Indoor walking differs from outdoor walking in that you are not traveling any appreciable distance. This means that the GPS in your iPhone cannot track the route you have traveled and calculate the distance you have covered. The Workout app nonetheless calculates the distance you have traveled, and if you have entered a distance goal, it will congratulate you when it determines that you have reached your goal. Lacking the additional information that GPS would have provided, the distance it displays will not be as accurate as it would be for outdoor walking, but it will be a reasonable approximation.

> **Note**   Speaking of approximations, the distances recorded for all the workouts that involve distance are approximations. Even though the display will give you your distance traveled, down to 1/100th of a mile or 1/100th of a kilometer, the accuracy is really not that good. You may find variations of as much as one-tenth of a mile on multiple traversals of the same course. For indoor walking and indoor running, the reading will typically be even less accurate. However, it will still be in the ballpark. If your goal in working out is to improve your overall fitness, putting in an exact number of miles per day is probably not the important thing. The important thing is that you are working out regularly at the level of effort that is appropriate for where you are in your conditioning program.

Indoor Running

You can get in an indoor running workout by cranking up the speed on your treadmill to your preferred running pace. Everything else remains the same as for indoor walking. The Workout app counts calories burned, tracks the duration of exercise, and approximates "distance" traveled; it then congratulates you when you reach the workout goal that you have set. You will burn calories faster while running than you would by walking, so you should reach your calorie goal sooner. Of course, if you set an exercise duration goal, that's how long the exercise will last, whether you ware walking or running.

Indoor Cycling

Well-equipped gyms today have stationary bicycles that you can "ride" to maintain fitness when riding a real bicycle outside is not feasible. Bad weather is one reason not to mount a real bicycle. Another is being in a place where cycling is not possible or not safe. When on vacation on a cruise ship, for example, cycling around the promenade deck is not permitted. Luckily, all cruise ships have gyms, and most are equipped with stationary bicycles. If you live in a city that experiences smog alerts, exercising indoors is probably a good idea on those days. This applies to walking and running as well as cycling. During pollen season, people with allergies or other respiratory problems would be well advised to favor indoor workouts. The Indoor Cycle activity in the Workout app operates the same way as the indoor walk and indoor run activities. You can set a calorie goal, a time goal, or no goal, before pressing the Start button. Calories burned, elapsed time, and heart rate will be continuously recorded. At the end of the exercise, you can choose to either save or discard the record of the session.

Elliptical

Like the indoor walk, run, and cycle, working out on the elliptical machine does not involve movement from one place to another, so GPS tracking is not effective. However, it is a real workout and can be quite strenuous. The extensive testing that Apple did before the release of the Apple Watch included taking measurements of the energy expended by a variety of people as they exercised on elliptical machines. Based on those tests, it calculated the number of calories expended, based on the elapsed time and the exerciser's heart rate. The watch display shows all three: elapsed time, calories burned, and heart rate. An elliptical workout can be a vigorous aerobic exercise that strengthens muscles as well as the cardiovascular system.

Rower

The rowing machine in the gym simulates the exercise your body would get if you were rowing a boat. It works your shoulders and your calves as well as your cardiovascular system. You can burn more calories on a rowing machine than you can on an elliptical machine, for the same amount of minutes spent on the exercise.

As is the case for the elliptical workout, the Apple Watch records elapsed time, calories burned, and heart rate while you are rowing. If you save the data taken during a workout, you can review it later. The app will retain both your longest and your most recent workout. You can set either one as a goal to shoot for that same duration again, a calorie burn goal, or a new goal of your choosing, or you can exercise without a specific goal.

Stair Stepper

The stair stepper is another machine you will find in health and fitness gyms, designed to build your cardiovascular fitness. As the name implies, it simulates the action of climbing a flight of stairs. For that reason it also helps to strengthen the muscles in your legs and lower body. As with the elliptical machine and rower, your Apple Watch will record and display your heart rate and will also record elapsed time and estimated calories burned. The calories-burned estimate, as is the case for all the other exercises that GPS cannot detect movement in, is based on tests run at Apple's fitness lab. You can set a goal for the duration of your workout, or you can record your session with no fixed goal.

Other

Of course, there are a lot of exercises you might do that are not included in the previous list. For these you can select the Other option. Your Watch will assume that whatever you are doing, it will work your body at a level equivalent to a brisk walk. As with the stationary exercises, elapsed time, calories burned, and heart rate will be recorded and will be saved if you choose to save them.

How the Workout App Works

The Workout app pulls in heart rate data from your watch's heart rate sensor, location data from your iPhone's GPS signal, and motion data from the watch's accelerometer. The accelerometer measures total body movement and steps taken. Its readings are used to calculate the calories you burn throughout the day. The heart rate sensor delivers data that is used to help calculate the intensity of a workout, which in turn feeds into the calculation of calories burned.

Setting Workout Goals

People come in all shapes and sizes, ages, and levels of conditioning, and they have different ideas about what they want to gain from an exercise program. Some just want to slow the decline in their physical capacity as they age. Others may want to build up their body to the point where it is performing at an elite level. In between there are a near infinite number of intermediate goals.

One of the first things you should do after consulting your doctor on any limits that you should not exceed is to set a fitness goal to work toward. If your goal is simply to slow the decline of functionality because of aging, starting an exercise program will surely do that. You will almost surely find that your goal is too conservative. A regular exercise program should give you increased strength and stamina, as long as you are consistent, giving your health the high priority it deserves.

A more ambitious goal might be to run in a 10-kilometer (10K) race at some point. Your goal should be ambitious enough such that you will have to stretch in order to accomplish it but not so difficult that it is truly beyond what you will ultimately be capable of. Once you do accomplish your initial goal, you can always set a higher one.

Building Cardiovascular Fitness

Your cardiovascular system, consisting of your heart, lungs, and all the blood vessels in your body, is more important for keeping you alive than anything else. As such, it should be the top priority of your workout program. All exercises will raise your heart rate above what it is when you are sitting quietly, reading, watching TV, or typing a book manuscript as I am doing now. However, the exercises that will do your cardiovascular system the most good are the ones that will raise your heart rate the most, as long as they don't put you in danger from overexertion.

The nine workouts featured by the Workout app are all good for building up your cardiovascular system. You may choose one that you particularly like, or you could mix things up and do one exercise on one day and another exercise on a different day. You could even do a different exercise on each day of the week. You would get your cardio work in that way, but you would also work different muscle groups, giving you a more all-around strength-building experience.

Shaping and Toning

Beyond the basic cardio workouts, some people may want to improve their appearance by creating a more classic shape. Abdominal exercises are an example of workouts that can help with this. You can use the Workout app when doing ab exercises, using the Other choice rather than one of the nine specific ones such as Outdoor Run or Elliptical. You may also want to use one of the third-party apps specifically targeted at ab workouts. I will cover these in Chapter 9.

Building Strength

Any exercise that you do will build strength in one or more muscle groups. Running will surely build strength in your leg muscles, as will cycling, providing you do it vigorously. Even walking slowly will at least help you keep the strength you already have. All these activities also strengthen your cardiovascular system. The elliptical machine and rower also work your arms and upper body. Using the Workout app along with these activities will help to keep you disciplined and will give you a sense of how much effort you have expended doing each of these exercises. While the Workout app is active, heart-rate data is being continuously sent to the Health app on your iPhone. You can review this data after a workout to get a sense of how hard each exercise is pushing you.

Tracking Progress with the iPhone Health App

The Health app on the iPhone acts as a repository for up-to-date information on the state of its owner's health. It accepts information from a variety of health-related apps, devices, and its paired Apple Watch to create a picture of the current health of the owner, as well as historical health data. Some of that data will be automatically routed to the Health app from the Workout app, some will be routed from third-party apps, and some can be entered manually by the iPhone owner.

The Dashboard

The Dashboard gives you a quick picture, in the form of line graphs, of how you have been performing. You can look at a year's worth of data, a month's worth, a week's worth, or a day's worth. The quantities being tracked are either data points sent in by apps or data points entered by you, based on measurements such as weight, which are not measured by a connected device. There are a number of different quantities that you can display on the Dashboard. I will briefly describe the ones that I find most useful. Figure 6-5 shows the Dashboard.

Figure 6-5. Health app Dashboard

Walking + Running Distance

Walking and running distance is tracked using your iPhone with its GPS capability. Software determines by the speed at which you are traveling whether you are either walking or running or traveling in some kind of vehicle, such as a bicycle or automobile. Only the distance covered at a normal walking or running pace will be recorded in the Walking + Running Distance category. If you are a bicyclist rather than a runner, your workouts will be recorded under Cycling Distance. Figure 6-6 shows the Walking + Running Distance graph with the current month's data displayed.

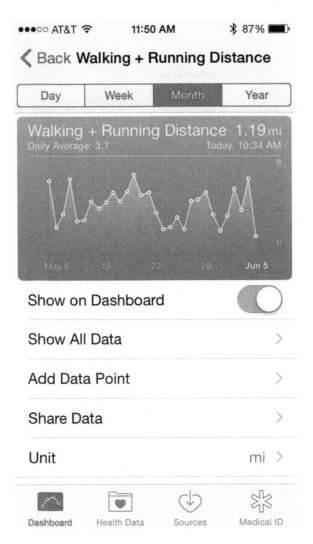

Figure 6-6. Walking + Running Distance display on Dashboard

There is an option to either show or not show this chart on the Dashboard, as well as an option to show all the data in this category that the Health app has collected. You can also add a data point manually.

Weight

Neither the Apple Watch nor the iPhone has a way of measuring your weight…yet. To keep track of this, you will need to enter your weight manually. Figure 6-7 shows how my weight has varied during the current month.

Figure 6-7. Weight display on Dashboard

Body Mass Index

Body mass index (BMI) is a simple computation based on a person's height and weight. Assuming your height doesn't change from day to day, it will track changes in weight. The Health app will do the computation for you. All you need to do is accept the answer it provides you. BMI moves up and down in sync with changes in weight, so the BMI graph looks exactly like the Weight graph.

Heart Rate

The Health app uses the heart-rate sensor in the Apple Watch or a sensor embedded in a chest strap paired with either the Workout app or a third-party fitness app to record your heart rate during an exercise. If your Workout app is running, it will take data from the Apple Watch. If you are wearing your chest strap and running a third-party fitness app, the Health app will take data from that. The Dashboard display shows a bar graph in which the bottom end of each day's bar shows the lowest heart rate during the day, and the top end shows your highest heart rate during the day. Figure 6-8 shows the current month's high and low heart-rate data.

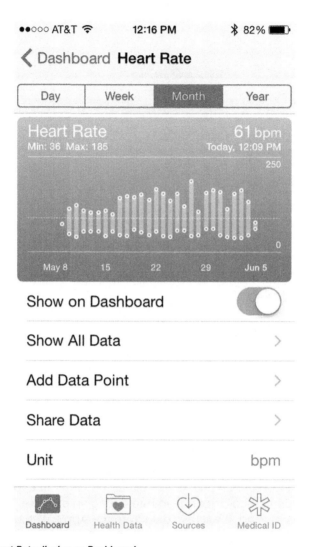

Figure 6-8. Heart Rate display on Dashboard

Steps

The accelerometer in your Apple Watch and the one in your iPhone emit a distinctive signal when you take a step, either while walking or running. The Health app counts these steps and displays them on the dashboard. One way to improve your fitness is to take more steps during the day. With the Steps display on the Dashboard, you can see how you are doing and perhaps reverse a trend of ever-lower step counts. Figure 6-9 shows the current month's data for Steps.

Figure 6-9. Steps display on Dashboard

Workouts

The Workouts display on the Dashboard records the time you spend exercising each day, where a level of exertion at or above that of a brisk walk counts as time spent working out. Workout data is recorded as a bar graph, and your average time spent working out over the past month is shown as well as the amount of time you spent working out yesterday. After your first workout of the day, yesterday's data is replaced by today's data.

Health Data

With the Health app on your iPhone, you can record and retain information about every aspect of your health and level of fitness. You can enter some information yourself, and other data will be sent automatically to the Health app from the Workout app on your Apple Watch or from third-party fitness apps running on your Apple Watch or iPhone.

Body Measurements

Body Measurements is an example of a category of information that currently you must enter manually. The items measured are as follows:

- Body fat percentage
- Body mass index
- Height
- Lean body mass
- Weight

Height and weight are easily measured, and the Health app will compute the body mass index from those two quantities. Body fat percentage and lean body mass require a more elaborate measurement. If you have been tested for those things, you can enter the results of the tests here. If you have not, in most cases you can do just fine without that data.

Fitness

The Fitness category tracks the things you have been doing to improve your overall fitness. In many cases these are the same things that you have been tracking on the Dashboard. The categories are as follows:

- Active Calories
- Cycling Distance
- Flights Climbed
- NikeFuel
- Resting Calories
- Steps
- Walking + Running Distance
- Workouts

When you use the Apple Watch Workout app, it will track active calories burned, cycling distance, resting calories, steps, walking and running distance, and workouts. NikeFuel is a program conducted by Nike (surprise!) that uses the Apple Watch to measure the amount of effort you put into movement. In that sense, it is similar to the Apple Watch's Workout app. If you use it, your results will show up in the Health app. Flights climbed is an entry you can make manually after climbing one or more flights of stairs.

Me

In the Me category, you can enter a little information about yourself.

- Birth date
- Sex
- Blood type

This data isn't all that valuable. Your birth date is handy so people can send you a cheerful birthday greeting once a year. Your sex is generally pretty evident just by looking at you. Your blood type might be handy in an emergency situation in which you need a blood transfusion. However, nobody is going to give you a transfusion based just on what your iPhone says. They will take a sample of your blood and type it before giving you a transfusion.

Nutrition

You would have to be a real diet weenie to fill in all the data that could potentially be recorded in this category. You would have to enter it all by hand, since neither your Apple Watch nor your iPhone can sense what you eat and drink. Among many others, nutrients you can enter include the following:

- Biotin
- Caffeine
- Calcium
- Carbohydrates
- Chloride
- Chromium
- Copper
- Dietary calories
- Dietary cholesterol
- Fiber

Good luck entering the number of micrograms of biotin and milligrams of caffeine you consume every day.

Results

Elite athletes preparing for major competitions want to get every last bit of performance out of their bodies, whether it's that last hundredth of a second in a sprint or that last kilometer in an ultramarathon. To do so, state-of-the-art training programs test and try to optimize every aspect of an athlete's body and mind. The Health app gives you the opportunity to record physical parameters that are indications of your state of health. Many of these require elaborate instrumentation, but if you have access to these resources, you may be able to squeeze some extra performance out of your body. Tests include the following:

- Blood Alcohol Content
- Blood Glucose
- Electrodermal Activity
- Forced Expiratory Volume, 1 sec
- Forced Vital Capacity
- Inhaler Usage
- Number of Times Fallen
- Oxygen Saturation
- Peak Expiratory Flow Rate
- Peripheral Perfusion Index

Some of this data will come in from source apps, and others can be added manually.

> **Note**   The Apple Watch uses its green LED pulse oximeter to measure heart rate. However, a pulse oximeter is a sophisticated instrument and is capable of measuring more than that. One thing that it could potentially measure is the peripheral perfusion index, which is a measure of the strength of a person's pulse. In its first release, the Apple Watch does not report this information. Possibly it will in a later release after Apple has had some time to perfect the technology.

Currently all the results in this category must be entered manually. Some will always require manual entry, but others may become measurable with later releases of the source apps that feed data into the Health app.

Sleep

With the Sleep category, you can record either the amount of time you spend in bed each day or the amount of time you are actually asleep. For "in bed" time, you can record the time you get into bed and then record the time you get up the next day. For "asleep" time, it is a little harder to make accurate entries. After you get into bed, how do you know how long it will take you to go to sleep? The next morning, how do you know when you stopped actually sleeping before being alert enough to make an entry in the Health app? Perhaps a rough estimate of these times is enough for most purposes.

Vitals

Your vital signs are the measurements that most closely track your state of health. If they fall out of the normal range, you have a serious health problem. When they flatline, you are dead. The following are stored by the Health app:

- Blood pressure
- Body temperature
- Heart rate
- Respiratory rate

Of these, heart rate is measured by the Apple Watch while the Workout app is running. It is also measured by some third-party apps that also serve as sources for the Health app. Blood pressure is a quantity that can be calculated, based on measurements made by a pulse oximeter. This function is not available in the initial release of the Apple Watch but may become available in a later release. Currently, there is no sensor for body temperature in the Apple Watch, and it has no way of measuring respiratory rate in breaths per minute. Whenever you have these variables tested, you can enter the results manually.

Sources

The Health app can draw data directly from physical devices, such as your Apple Watch and your iPhone, as well as from third-party apps. The following are a few apps that can serve as sources for the Health app along with the types of data they share.

MyHeart

MyHeart is one of the Research Kit apps. Run by Stanford University, it monitors study participants for possible cardiovascular abnormalities. MyHeart will share only two data items with the Health app: a person's height and weight. However, it will gladly accept many data items that the Health app is willing to share with it, such as the following:

- Blood glucose
- Cycling distance
- Date of birth
- Diastolic blood pressure
- Flights climbed
- Heart rate
- Height
- Oxygen saturation
- Sex
- Sleep analysis
- Steps
- Systolic blood pressure
- Walking and running distance
- Weight
- Workouts

By taking and analyzing all this data, the researchers hope to determine what factors increase chances of contracting cardiovascular disease and what practices or characteristics might be protective against it. Some of these data items, such as systolic and diastolic blood pressure, cycling distance, steps, and walking and running distance can be gleaned from the Apple Watch Workout app. People enrolled in the MyHeart program can contribute data to the program every time they engage in a workout.

Note   Research Kit is a collaboration between Apple and several groups of academic researchers, investigating various diseases and conditions. The idea is to collect data from test subjects who own an Apple Watch and are willing to share their data with the researchers.

RunKeeper

The RunKeeper fitness app shares more than just height and weight information with the Health app. It also shares active calories burned, cycling distance, walking and running distance, and workouts.

The Health app will share all those same things, in the reverse direction, with RunKeeper as well as sharing the date of birth and sex of the user.

Runtastic and Runtastic Pro

Runtastic, like RunKeeper, shares active calories burned, cycling distance, walking and running distance, and workouts with the Health app but does not share height and weight data. Furthermore, Runtastic does not accept any information coming in the reverse direction from the Health app. If you want any data to be recorded in either Runtastic or Runtastic Pro, you will have to enter it manually rather than it coming in from another app.

Strava

Strava is another fitness app similar to RunKeeper and Runtastic. Like those two apps, it will write to the Health app the data that it takes for active calories, cycling distance, walking and running distance, and workouts. It will also read from the Health app the date of birth, sex, and weight information.

Aside from RunKeeper, Runtastic, and Strava, there are many other fitness apps, too numerous to mention, that make use of the sensors in the Apple Watch. Try several and then stick with the one you like best.

Medical ID

The Medical ID screen in the Health app gives your name and birth date as well as any medical conditions that you have chosen to reveal about yourself. If emergency medical personnel encounter you at a time when you are unable to communicate, information listed here could enable them to give you more appropriate treatment than they otherwise could.

Setting New Goals

After you have set goals that stretch your capabilities just a little bit, you will eventually reach a point where you are able to meet those goals consistently. When this happens, it is time to set new goals. Perhaps you should be running a longer distance or at a faster speed. Perhaps you should increase the number of minutes per day that you spend working out. Aim to always be in the position of being just slightly out of your comfort zone. A workout should be hard work. Just make sure you do not overdo it.

Summary

The Workout app is the Apple Watch's primary tool for motivating you to engage in vigorous exercise of some form every day. If you use it on a consistent basis, you *will* get stronger and faster and have more endurance. The secret to progress is to use the app consistently. Once use of the Workout app becomes habitual, your fitness level will improve as a matter of course.

Chapter **7**

Using the Apple Watch with the iPhone Health App

Strictly speaking, the Health app is an iPhone app, not an Apple Watch app. It was around long before the Apple Watch was introduced. However, the arrival of the watch has increased the value of the Health app tremendously. The Apple Watch senses your movements and vital signs in real time while you are exercising and transmits that information directly to the Health app via the Bluetooth connection between your Apple Watch and your iPhone. The Health app records this data and presents it to you in graphical form as well as in the form of raw data. With the charts on the Health app Dashboard, you can tell at a glance how far you have run or walked today, what the range of your heart rate has been today as well as for the current month, and perhaps the number of steps you have taken.

Configuring the iPhone Health App

iOS8 and newer versions of the Apple operating system on iPhone will have an Apple Watch icon on the Home screen, as shown in Figure 7-1.

Figure 7-1. *iPhone screen showing Apple Watch icon*

When you tap the icon, the My Watch screen shown in Figure 7-2 appears.

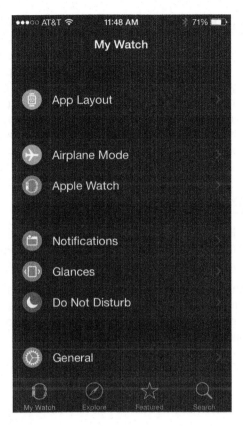

Figure 7-2. *iPhone My Watch screen*

The My Watch screen first shows controls such as App Layout, Airplane Mode, and Apple Watch, followed by a list of built-in apps. Scroll down to find the Health app. Select it and enter the information that it asks for, which includes birth date, sex, height, and weight. These facts will be used by formulas that calculate key information, such as your body mass index and how hard you are pushing yourself during an exercise. This information may also be shared with third-party fitness apps to augment the information that they collect on your workout sessions. One of the cool things about the Health app is that it acts as a central hub for all your health and fitness apps, both native and third party. They can all share data with the Health app, and vice versa.

Setting Your Watch as a Data Source

Once your Apple Watch has been paired with your iPhone and once the built-in apps, such as the Activity app and the Workout app, have been downloaded to it, your Apple Watch will automatically be considered as a data source for those apps. You will also want your watch to be a data source for third-party health and fitness apps that you may have. I will discuss several of these apps in Chapter 9, along with how to set your watch as a data source for them.

Using the Health App Main Menu

When you tap the Health app icon to call it up, you will be placed into one of four modes, identified by icons at the bottom of your iPhone screen. They are Dashboard, Health Data, Sources, and Medical ID.

The Dashboard, shown in Figure 7-3, displays graphs of metrics showing a history of your fitness and health stats. Some of these are provided by default to get you started; others you can add yourself. You can also delete default graphs that you don't want.

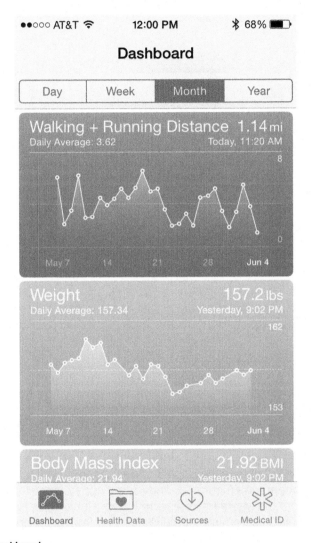

. *Figure 7-3. Dashboard*

The Health Data screen, shown in Figure 7-4, is divided into seven major categories. Many of these hold data that you enter manually, but two of them, Fitness and Vitals, also can accept data directly from the Apple Watch, as well as from third-party applications.

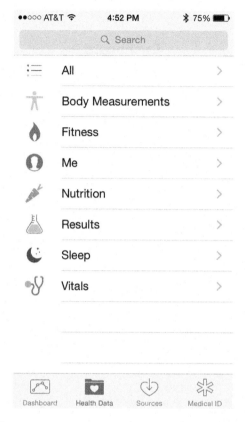

Figure 7-4. Health data

The Sources screen shows the apps and devices from which the Health app can accept data. As Figure 7-5 shows, my copy of the Health app can accept data from the MyHeart ResearchKit app, Runkeeper, Runtastic Pro, and Strava, as well as from my Apple Watch and from my iPhone.

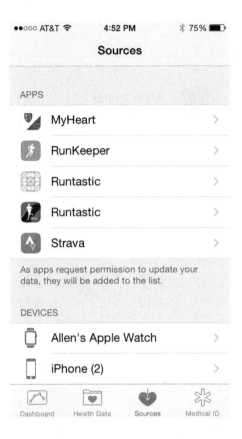

Figure 7-5. Sources

The Medical ID screen (Figure 7-6) shows your picture, your name, and your age, as well as any medical conditions that might be helpful if you ever need medical help but are not able to communicate.

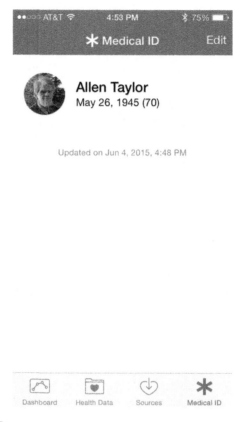

Figure 7-6. Medical ID

Tracking the Data That Interests You

The data-gathering capabilities of the Apple Watch can be both a blessing and a curse. They are a blessing in that you are able to record and analyze information that it was not practical to collect in the past. They are potentially a curse in that so much data is collected that separating the signal from the noise can be daunting.

What you want to do is track the data that interests you and be able to easily ignore the rest. For tracking the data of interest, the Health app's Dashboard graphs give you an overview of the data items of interest and show such things as variability from day to day as well as trends in the data.

Data is captured and maintained in these seven categories:

- Body Measurements
- Fitness

▓ Me

▓ Nutrition

▓ Results

▓ Sleep

▓ Vitals

All of these areas are important to a person's overall health, but the Apple Watch is particularly involved with the Fitness and Vitals categories.

Fitness Data

The main message that the Apple Watch conveys regarding fitness is that action enhances fitness. The more active you are, the more fit you are likely to become and likely to remain. There are a variety of things you can do to remain active. Having multiple activities that keep you moving will also keep things interesting. With its accelerometer and with the iPhone's GPS, the Apple Watch can get a good sense of how active you are. Primarily, it measures activity in terms of calories burned.

Active Calories

Calories (actually kilocalories but called *calories* for short) are a measure of heat. Just being alive and keeping your body up to 98.6 degrees Fahrenheit burns calories. Even sitting still burns calories. Sitting motionless while contemplating your next move in a tough chess game burns more calories than you would imagine. It is said that the brain actually uses up to 20 percent of the total calories consumed by a person. What those chess players claim is actually true: thinking is hard work.

However, the calories you burn keeping alive and thinking are not active calories. Active calories are the calories that you burn when you go above and beyond and get your body moving. The more vigorously you move, the more active calories you will burn.

Cycling Distance

The Workout app discussed in Chapter 6 enables you to record both outdoor and indoor cycling workouts. When you are cycling outdoors, the GPS function on your iPhone will enable the app to calculate the distance you have traveled. When you are cycling indoors on a stationary bike, the Workout app will estimate how far you have gone, based on accelerometer readings and heart rate. In both cases, the Workout app will send its calculated distance to the Health app, and it will show up on the Health app's Dashboard as well as appearing as numeric data when you tap the Show All Data option of the Fitness ➤ Cycling Distance category.

Flights Climbed

The Workout app does not record actual flights of stairs climbed but does record workouts on a stair-stepper machine in the gym, which simulates climbing flights of stairs. If you want to record actual flights of stairs that you have climbed, you can enter the number of flights manually and keep track that way. This is the same thing you would do, for example, to manually enter your weight after you weigh yourself on your bathroom scale every day.

If you want to record time spent on a stair stepper, you will have to translate the time you spend on the machine into flights of stairs climbed and then enter that number manually into the Health app's Health Data section, namely, the Fitness ➤ Flights Climbed category.

NikeFuel

NikeFuel is a metric for measuring whole-body movement. It is implemented with an app running on either an iPhone or Android phone, which uses accelerometer readings to sense and record your body's movement. Movement that NikeFuel detects is routed to the iPhone Health app, where it can be displayed on the Dashboard along with the other fitness metrics that you choose to display.

Resting Calories

The Health app differentiates between active calories and resting calories. Resting calories are burned by keeping your heart pumping and other vital processes operating, as well as contemplating your next chess move. You also burn a few resting calories while deciding whether you want your groceries to be bagged in paper or plastic. The Health app records resting calories, but neither the Activity app nor the Workout app includes resting calories in their totals. If you want to record resting calories, you will have to do so manually, possibly using figures you obtain from a third-party fitness app that records total calories rather than just active calories.

Steps

The Apple Watch uses its accelerometer to estimate the number of steps you take in the course of a day. Those readings are automatically transferred to the Health app on your iPhone and show up on the Dashboard's Steps graph. The graph gives you a rough idea of how active you are by showing how many steps you have taken so far today, compared to how many you took each day of the current week, month, or year. Figure 7-7 shows the Steps graph and the controls you can use to configure the display.

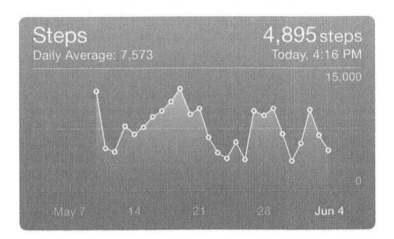

Figure 7-7. Steps graph

The graph shows you how many of your steps your source app or device has recorded so far today, as well as a history of how many steps you have taken in past days.

Walking + Running Distance

For most people, walking and running are the two activities that will make up the bulk of a person's body movement activity, although cycling may rate high for some. Walking and running recorded either by the Workout app or by one of the third-party fitness apps that runs on the watch will be transferred to the Health app and appear on the Walking + Running Distance graph on the Dashboard. A quick glance at the graph (Figure 7-8) will tell you whether you need to go for a run before you wrap things up for the day.

Figure 7-8. Walking + Running Distance graph

Workouts

The Workouts graph (Figure 7-9) shows the time spent working out per day in hours and minutes. Data comes from the Apple Watch's Workout app and also from third-party fitness apps.

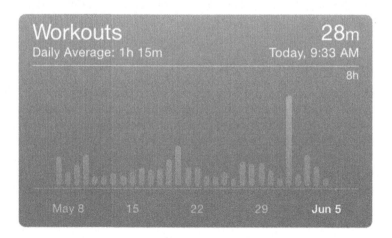

Figure 7-9. Workouts graph

> **Warning**   If you activate both your Watch's Workout app and a third-party
> fitness app, the Workouts graph will take data from both, making it look like you
> are working out twice as long as you really are. Your workout is counted twice.

This just means that if you are concerned about accurately tracking your workout time, you should choose to use either the Workout app or one third-party app of your choice. Having, for example, Runtastic and Strava both running while exercising would also double count your workout time.

Vitals Data

Under the Vitals category, the basic signs of your overall level of health are recorded. These are the things that health professionals check whenever you go to the doctor for any reason: blood pressure, body temperature, heart rate, and respiratory rate. If any one of these indicators deviates from what is considered normal, that gives the doctor a clue as to where your trouble might lie. Of course, it's good to know this information even before you go to the doctor. Early detection of a problem could cause you to seek treatment before a condition becomes acute.

Blood Pressure

Although the Apple Watch does contain a pulse oximeter, which theoretically can be used to measure blood pressure, blood pressure readings are not a feature of the watch's first release. Speculation abounds on why this may be so. However, since the hardware is present, it is possible that a future software update may release the capability to measure blood pressure. Then again, maybe not. We will just have to wait and see.

Body Temperature

When you visit the doctor's office, a medical assistant probably sticks a temperature probe in your mouth to take your temperature. The Apple Watch does not have a temperature probe, and even if it did, you probably would not want to stick it into your mouth anyway. Any time you do have your temperature measured, you can manually add that measurement as a data point to the Health app if you want.

Heart Rate

The Apple Watch measures your heart rate, and those measurements show up in the Health app (Figure 7-10). This happens automatically, without any manual intervention. Readings are taken throughout the day, and a bar on the Heart Rate graph extends from the lowest heart rate measured during the day to the highest. If the automatic measurements do not happen as often as you like, you can always start the Workout app, which will take readings on a frequent basis.

Figure 7-10. Heart rate graph

If you want, you can also see the raw data that is used as the basis for the graph (Figure 7-11).

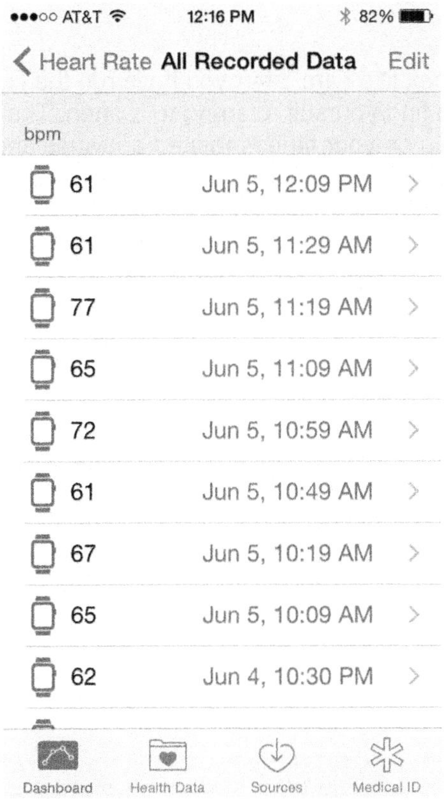

Figure 7-11. *Showing all data*

Respiratory Rate

Respiratory rate is measured in breaths per minute. Like blood pressure, this is a vital sign that could be measured by the sensors within the Apple Watch but is not offered in the first release of the Apple Watch. Picking up rhythmic chest expansions from the wrist is a challenging measurement to make. We will have to wait to see whether this is ever offered. However, the capacity is there to record this data in the Health app, so if you ever have your respiratory rate measured at a clinic or sports medicine facility, you can enter that data manually, and it will appear on the Health app Dashboard.

Summary

If you are a stats geek like I am, after you have run the Health app for a while, you will soon find yourself referring to it often. The information it records and displays on your fitness-related activities and heart rate gives you a near real-time view of the state of your fitness and your overall health. It can also give you advance warning of incipient problems before you would otherwise become aware of them. The usefulness of the Health app is multiplied by its connection to the Apple Watch, over what it is when used only with the iPhone.

Chapter 8

The ResearchKit Health Projects

A physician's guiding principle is "First, do no harm." As a result, new treatments for diseases must be tested thoroughly, preferably in double-blind, placebo-controlled, randomized clinical trials, before they can be approved and then prescribed to patients. For trial results to be considered statistically significant, a large number of study participants are needed, and typically recruiting enough participants with a given disease or with susceptibility to the disease is difficult. Progress in translating research into clinical treatments is slower than it could be because of this bottleneck in the process of conducting clinical trials.

The recent phenomenon of crowd sourcing has the potential to break the logjam by opening up a large pool of potential test subjects for research into common ailments. Millions of people own iPhones and Apple Watches. Many of these people would like to see medical research move ahead faster and are willing to devote some of their time and effort to help. Apple's ResearchKit is aimed at those public-minded Apple Watch and iOS device owners.

What ResearchKit Is and How Researchers Can Use It

ResearchKit is a platform upon which applications can be built that gather information from the sensors on the Apple Watch and iPhone, which specifically address highly targeted health-related questions. Researchers in a wide variety of medical fields can team up with app developers to create apps that monitor the location, activity, and vital signs of participants on a continuous basis.

With millions of Apple Watches on the wrists of people all over the world, researchers will have little difficulty recruiting thousands of participants. The large number of enrolled participants massively increases the statistical power of the conclusions that are drawn from the research studies. The research can be conducted much more cheaply than would normally be the case, an important point at a time when research grants are getting harder to win.

Diseases Initially Addressed by ResearchKit Apps

The first research studies to use ResearchKit apps are designed to gain insight into five diseases that cut a path through huge swaths of the populace, causing great suffering and drastically shortening lives. Let's take a brief look at each.

Asthma

Asthma is a respiratory disease due to chronic inflammation of the airways in the lungs. It is characterized by wheezing, shortness of breath, chest tightness, and coughing. In severe cases, it can be fatal. Approximately 300,000 people die per year worldwide from asthma, and about 280 million people are affected by it to a greater or lesser degree. Asthma is the third leading cause of death in the United States after heart disease and lung cancer.

Asthma is caused by a combination of genetic and environmental factors. There are a number of genes that are involved and a wide variety of environmental conditions that can trigger a severe asthma attack among those who are genetically susceptible.

Mt. Sinai, Weill Cornell Medical College, and LifeMap have developed the Asthma Health app with ResearchKit. In addition to collecting data daily on the condition of study participants, using the Apple Watch link to its paired iPhone's GPS capability, it can also warn them of a potential asthma attack trigger, such as high air pollution when they are about to enter an area that is currently under an air pollution alert. Maybe it would be better to wait and go downtown on another day.

Parkinson's Disease

Parkinson's disease is a degenerative disorder of the central nervous system. It is characterized by movement-related symptoms including shaking, rigidity, slowness of movement, and difficulty walking. It is a progressive disease and in advanced stages affects cognitive function and behavior. Most cases occur after age 50, and the disease becomes progressively more prevalent at higher ages.

Globally, about 7 million people are affected by Parkinson's disease, of which about 1 million reside in the United States. In 2013, approximately 103,000 people died of Parkinson's disease worldwide. The disease is a tremendous and growing drain on the healthcare systems of all the countries of the world.

The University of Rochester and Sage Bionetworks have developed the mPower app, which uses the sensors and computing power in the iPhone and Apple Watch to measure a person's dexterity, balance, memory, and gait. The gyroscope in the Apple Watch can sense abnormal movements, enabling researchers to track the progression of the disease. Participants may also notice their own symptoms and report them.

Diabetes

There are several types of diabetes, the most prevalent of which is diabetes mellitus, also known as Type 2 diabetes. It is a metabolic disorder, characterized by high levels of glucose in the blood, due to insulin resistance. Incidence of diabetes mellitus has grown dramatically coincident with the worldwide growth of obesity in recent decades. As of 2010, approximately 285 million people had been diagnosed with the disease. Diabetes mellitus is a chronic disease, which typically chops ten years off the life expectancy of people who have the condition. People with diabetes are more likely to suffer complications such as heart disease, stroke, diabetic retinopathy, and kidney failure. Blood flow to the limbs is impaired in advanced cases, sometimes requiring amputation.

The Massachusetts General Hospital has developed the GlucoSuccess app, which is used to track the diet, physical activity, and medication use of study participants. It can also inform participants how their food choices, activity level, and medication compliance affect their blood glucose level. This could lead to behavioral changes that improve the quality and length of the participant's life.

Breast Cancer

Breast cancer is the most common invasive cancer in women. It affects men too but is 100 times more common in women than in men. Although treatment has improved dramatically in recent years and extended the lives of breast cancer survivors, there were still more than 521,000 breast cancer deaths worldwide in 2012. North America has the highest breast cancer rate in the world, so a disproportionate number of people with the disease reside in North America.

A diagnosis and treatment of breast cancer probably has a more devastating psychological impact on those who are diagnosed with it than on those with any other type of cancer. Depressed mood and fatigue are common side effects of treatment.

The Share the Journey app, developed by the Dana-Farber Cancer Institute, UCLA Fielding School of Public Health, Penn Medicine, and Sage Bionetworks enables study participants to give information about energy levels, cognitive ability, and mood. The object of the study is to find ways to improve the quality of life of people after they have been treated for their cancer.

Cardiovascular Disease

Cardiovascular disease is the number-one killer in the developed world. It shortens lives and causes great distress for those affected by it, either directly or indirectly. The *cardio* part of the word *cardiovascular* has to do with the heart, and the *vascular* part has to do with the arteries and veins that the heart pumps the blood through in order to supply oxygen to all the cells of the body. If a cell doesn't get oxygen, it dies. If enough cells die, the person dies.

Sometimes the blood cannot get to where it needs to go because blood vessels are clogged up with plaque. Other times a blood clot blocks a major vessel also preventing blood from going where it is needed. If the heart is not beating with a regular rhythm, that can also be deadly, as can various defects in the heart's valves.

The MyHeart Counts ResearchKit application, created by Stanford University in association with the American Heart Association, first asks participants a series of questions about themselves. These include the usual statistics, such as birth date, sex, height, and weight. It also asks what time a person usually goes to sleep and when they usually wake up.

After people sign up, they are asked to check in every day and answer a few questions. Some of these questions are asked only once, while those that might change on a daily basis are asked every day. In addition to the questions answered, the app also takes data recorded by the watch's gyroscope and accelerometer, as well as the iPhone's GPS readings, to tell how active you are.

Periodically, the app will ask you to take the six-minute walk test. For this test you are asked to walk as far as you can in six minutes. Over the course of the study, any change in your maximum walking speed will be noted. In addition to the distance walked, your average and maximum heart rate during the test will also be recorded.

Every day, the app will ask you whether you wore your Apple Watch all day and all night yesterday, or just all day. Since the watch's battery does not last 24 hours between charges, you will most likely answer "All day but not all night." You may also answer "About half the time" or "Rarely if at all." Figure 8-1 shows this screen.

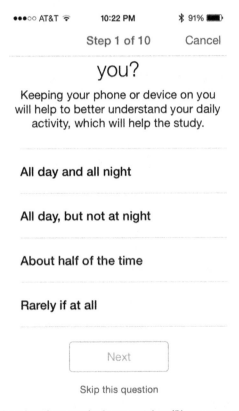

Figure 8-1. Question about how long you had your watch or iPhone on you yesterday

It will also ask you whether you engaged in any activities yesterday that were not recorded (Figure 8-2). If you say yes, it will ask how long and how vigorous those activities were. Finally, it will also ask you how many hours of sleep you logged last night (Figure 8-3).

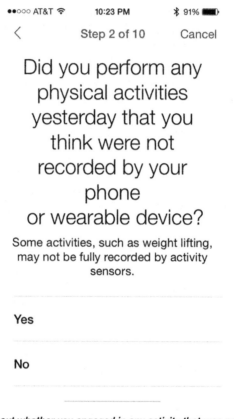

Figure 8-2. Question about whether you engaged in any activity that was not recorded

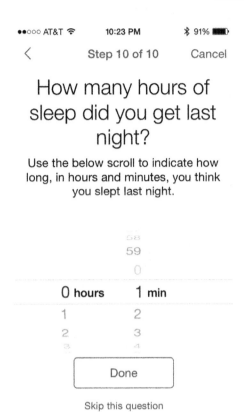

Figure 8-3. Question about how much sleep you got last night

What Future ResearchKit Apps Will Do

As time goes on, you can expect that many more researchers will use ResearchKit apps as a means of tapping into a large population of potential test subjects. If one of these is on a subject that you would like to help advance, you can volunteer to participate. Ideally, the result will be a benefit to many people in the future and possibly even yourself.

Summary

This chapter covers the ResearchKit tool that researchers in a variety of fields can use to gather data from large numbers of Apple Watch users, thus increasing the statistical power of any conclusions they come to. The first applications are all medical in nature, but any research that can make use of the physiological parameters or movement data recorded by the Apple Watch is a candidate for a ResearchKit application.

Chapter **9**

Third-Party Health and Fitness Apps

Although the built-in health and fitness apps are well designed (as we have all come to expect from Apple), they do not have the depth that is available from developers whose whole business revolves around health and fitness. Many mature health and fitness apps are available for the iPhone. Some of the best of these have extended their iPhone apps to make use of the convenience and added functionality of the Apple Watch. In this chapter, I will give brief descriptions of a few of them, highlighting the best features of each and describing the added capability provided by the watch.

In the first release of the software development kit (SDK), Apple did not give third-party developers the ability to acquire data from the watch's heart-rate sensor. With the advent of watchOS 2, this capability has now become available to developers. Alternatively, you can feed heart-rate data to your third-party apps by pairing a Bluetooth chest-strap heart-rate monitor with your iPhone. With the chest-strap monitor, you will not be able to see your heart rate on your wrist while exercising, but you may be able to see it by glancing at your phone, both in your third-party app and in Apple's Health app. The Health app will take heart-rate data from a Bluetooth chest-strap heart-rate monitor if one is present and from the Apple Watch if one is not present.

The purveyors of third-party apps all want to make money from their apps, but different apps will do that in different ways. Some apps will provide basic functionality for free and additional functions for premium or elite members, who pay a fee for the added value. Other apps, such as Nike+ Running, just want to encourage you to run more so that you will wear out your shoes sooner and buy a new pair of Nike shoes. In fact, Nike records not only the mileage of your runs but also the total accumulated mileage on your running shoes. When mileage exceeds a certain level, the app encourages you to buy a new pair of shoes. There is a direct link to the Nike store where you can buy not only shoes but all kinds of running apparel. Lacking an apparel store of its own, Endomondo runs advertisements for Under Armor and other sponsors at the bottom of the iPhone screen. The free version of Runtastic runs advertisements and also makes sure you know how to go to the Runtastic online store.

> **Warning**   A low-power Bluetooth connection between the iPhone and the Apple Watch, or between the iPhone and an external device such as a heart-rate monitor, can be a source of problems. Functions that once worked stop working. Apps that work on the iPhone don't run on the watch. In many cases, these problems are because of Bluetooth communication issues. They can often be resolved by deleting the third-party app having the problem and then reinstalling it.

Nike+ Running

Nike, as you might guess from the name of its app, is primarily interested in getting people *running*. Runners need good shoes, and they run through them (pun intended) quite often. Runners need to replace their shoes more often than, for example, weight lifters or cyclists. Furthermore, runners who are trying to improve their performance are sure to want the kind of statistical information that can be provided by the sensor-laden iPhone and the equally high-tech Apple Watch.

The iPhone Part of Nike+ Running

The Nike+ Running app was available before the Apple Watch was released and gives runners a lot of information about their runs but, like all smartphone apps, is not particularly convenient to interact with while a person is actually running. The main advantage for runners, or exercisers of any kind, of the Apple Watch is that you can garner useful information from it with a quick raise of your wrist and a glance at your watch.

The main menu of the iPhone part of the Nike+ Running app, shown in Figure 9-1, indicates that there is some real depth to this app, with a variety of categories. Let's take a brief look at them.

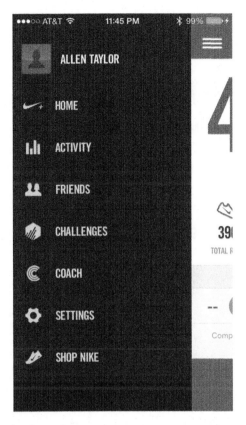

Figure 9-1. Nike+ Running app main menu

Home

The Home screen, shown in Figure 9-2, encapsulates your cumulative statistics. At the top it shows how far you have run since you started using Nike+ Running. In addition, it shows your average pace over that distance as well as the distance of your most recent run. Tapping BEGIN RUN at the bottom takes you to the Run Setup screen, where you can set up such things as whether you want to hear music while you run, whether you will be running outdoors or indoors, whether you want your phone display to be vertical or horizontal, whether you want to let your Facebook friends know that you have just started running, and whether you want to change your run settings.

Figure 9-2. Home screen

Note   If you are running indoors, for example on a treadmill, Nike+ Running cannot track your route with GPS and must estimate the distance you have run based on accelerometer and gyroscope tracking of your rhythmic body movements. This estimate will not be as accurate as the distance estimate of an outdoor run when you are carrying your iPhone with its GPS capability.

At the bottom of the Run Setup screen is the START button. Press it to begin a workout. A countdown timer that you set to a convenient interval, for example three seconds, begins. When it reaches zero, your run timer begins, as shown in Figure 9-3.

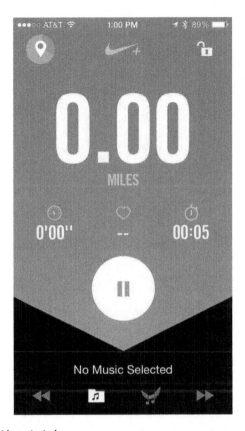

Figure 9-3. A workout has started

There is a Pause button that you can press if you want to interrupt your workout for any reason. It takes you to a screen (Figure 9-4) where you can either resume the workout or end it.

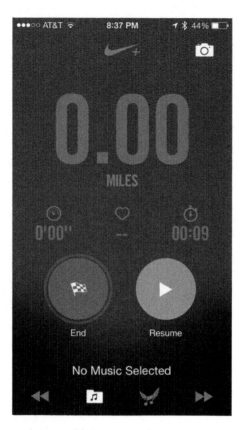

Figure 9-4. You have paused your workout

To resume the workout, all you need to do is tap the green Resume button. To end the workout, press and hold the red End button. The workout will end, and you will be returned to the Home screen, which will now show your updated stats.

Activity

The Activity screen shows you a synopsis of your most recent runs: your distance, your pace, and your elapsed time, as well as how you felt and the type of surface you ran on. By scrolling up, you can see records of earlier runs. By tapping the right edge of one of those runs, a map of the route will be displayed, and the color of your track will indicate your relative speed: green for fastest, yellow for slower, orange for even slower, and red for stopped. Figure 9-5 shows where I walked with my dog, Mojave, who liked to stop and sniff the ground every now and then. The times that he did show up as red spots. An emoji shows that I felt good on the walk, and an icon shows that I have less than three miles on my Gel Nimbus 16 shoes.

Figure 9-5. Detail of a walking workout

Friends

Nike+ is a social community. If you want to join the community, select Friends and then press the MAKE MY ACCOUNT SOCIAL button. You can request that people you know who are already social members of Nike+ accept you as a Nike+ friend. You can keep up on how active your friends are, and you can motivate each other to be more active.

Challenges

You can challenge any of your Nike+ friends to a race. You set the distance and the time limit. An example might be 30 miles within 30 days. The first person to run the required distance within the stated time wins a virtual medal. Everyone who completes the challenge wins something. You can also accept a challenge posted by one of your friends. This is a good way to engage with your community and to motivate yourself and others to be more active.

Coach

This option gives you a virtual coach who will prepare you for a race in the future. You can specify a 5K, 10K, half-marathon, or marathon, and the coach will give you a program that will take you through the conditioning you will need to complete such a race. The program will include walking, running, resting, and cross-training, getting progressively more intense as race day approaches.

Settings

This option gives you the opportunity to set a number of variables, such as privacy settings, to the way you want them. For our purposes, the most important thing here is the instruction on how to configure your iPhone so that it links to your Apple Watch so you can view your progress while running, with a glance at your watch.

Shop Nike

Nike would be missing a chance if it didn't take the opportunity to show you some of its featured products for runners, and this is the place it does it. The Shop Nike option is a direct link to the Nike.com web site.

Starting and Ending a Run from Your iPhone

You can both start and end a run with your iPhone as described earlier in the Home section. You can also start and end one from your Apple Watch. Let's look at that next.

The Apple Watch Part of Nike+ Running

Because of the small screen on the Apple Watch, there is not much room for displaying running statistics. However, the watch comes in handy by enabling you to both start and stop recording your run without hassling with your phone. When you select the Nike+ Running app from your watch's Home screen, you see the Start screen shown in Figure 9-6.

Figure 9-6. Nike+ Running Start screen

Tapping the figure of the runner launches the app, the audio coach on your phone announces "Beginning workout," and the run timer starts counting elapsed time.

Tapping the watch during the run rotates through displays of Pace, Duration, and Miles (assuming you did not set your app to measure distances in kilometers). Figure 9-7 shows the Duration display.

Figure 9-7. The elapsed time of your run is one thing that is displayed

A glance at your watch will let you know how you are doing while you are running.

Using Force Touch shows the display in Figure 9-8.

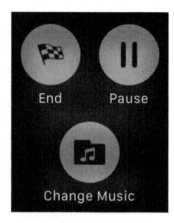

Figure 9-8. The End, Pause, or Change Music screen

At the end of your run, be sure to press the End button so that the statistics recorded are accurate. If you want to pause in the middle of a run for any reason, such as taking a phone call, press the Pause button. Since you can store more than one choice of music to run by on your watch, you can even start listening to a new mix by pressing the Change Music button. There is no way to plug earbuds or headphones into the Apple Watch, but you can access the music stored on it with Bluetooth headphones.

Endomondo

Although fitness apps such as Nike+, Endomondo, RunKeeper, Runtastic, and their competitors were originally designed to track running sessions and still work best for that sport, it is possible to use some of these apps for other sports too. Different sports put different demands on the body, so their calorie burn rates will vary. Many third-party apps, Endomondo included, calculate calorie burn differently, depending on which exercise you tell it you are doing. Running burns more calories per hour than walking does, for example.

The iPhone Part of Endomondo

Like Nike+ Running, Endomondo has a main menu of options, shown in Figure 9-9.

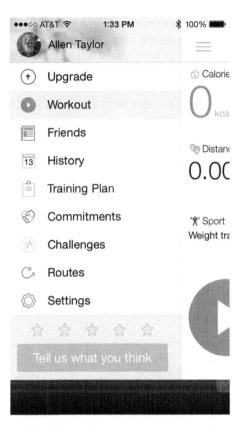

Figure 9-9. Endomondo main menu

Let's look at each option in turn.

Upgrade

Most of these third-party apps work on the same revenue model. A lot of effort and expense goes into developing these apps, and the companies that create them want to recover their investment and make a profit on top of that. They do this by providing a basic set of features for free and then charging for a "premium" or "elite" or perhaps "gold" membership, which unlocks additional features that the user may find to be of value. After a user of the free product has developed a history with that product, she is unlikely to switch to a competitor and lose all that historical data about her past performance and progress. As she becomes more dedicated, she is more likely to upgrade.

In Endomondo's case, a premium membership will give you additional statistics, access to training plans, and analysis of heart-rate monitor data. Because first-generation third-party apps do not have access to the Apple Watch's heart-rate data, you must use a chest-strap monitor with a Bluetooth connection to feed heart-rate data to Endomondo. When the Endomondo app is updated to work with watchOS 2, check to see what additional capabilities are present.

Endomondo helps support the free version of the app with advertisements. If you buy the premium program, you can elect to replace the space the ads take up with data that you might actually care to see.

There are a number of additional features available to premium users that people might find worth the cost of a membership.

Workout

During a workout, the app displays the number of kilocalories burned so far, the distance traveled, and the user's heart rate in beats per minute. Figure 9-10 shows the workout screen.

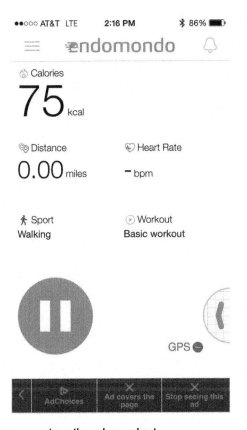

Figure 9-10. Workout screen partway through a workout

This figure was captured partway through a walking workout when no heart-rate monitor strap was being worn and no GPS signal was being received. Lacking GPS, the app could not record the distance traveled.

> **Note**   Some fitness apps, such as Nike+ Running, will estimate distance traveled using body movements when GPS is not available, but Endomondo does not do this; it displays 0.00 miles.

Figure 9-11 shows the result screen at the end of a workout when GPS is not working.

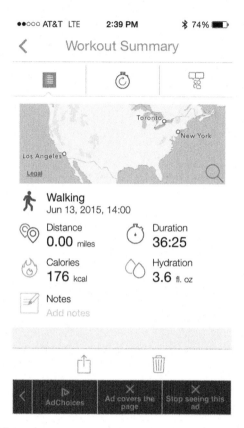

Figure 9-11. Workout Summary screen

In addition to date and time of the workout, duration in minutes, calories in kilocalories, and hydration in fluid ounces are recorded. For a non-GPS workout, Distance shows as 0.00 miles. You can add a note if you want. The map at the top shows the entire contiguous United States because, lacking GPS, Endomondo could not narrow down location more precisely. There is an advertisement for Endomondo Premium at the bottom of the screen, which does not show in Figure 9-11. Nike+ Running does not display ads on the free version of the app but reminds users when they should start thinking about replacing shoes that have put in a lot of miles, and the main menu contains a link to the online Nike store.

Friends

The Friends option has two functions: it records your activities and shares them with friends, and it shares your friends' activities with you. Figure 9-12 shows the screen displaying my most recent activities.

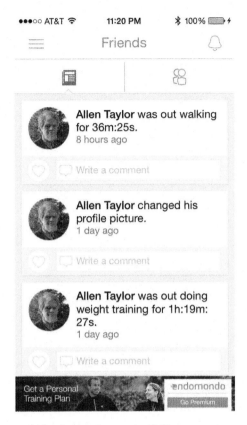

Figure 9-12. Friends screen, showing user's recent activities

Tapping the friends icon at the top right will switch the display to show the recent activity of your friends who are also using Endomondo. Endomondo friends are friends by mutual agreement. One of you must send an invitation, and the other must accept it.

History

When you select the History menu item, a scrollable screen appears that lists pointers to workout summaries of all the workouts that you have done since you started using Endomondo. Figure 9-13 shows my most recent workout history.

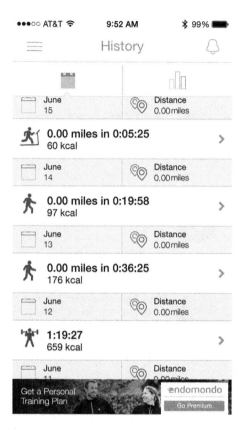

Figure 9-13. Recent workouts

In addition to today's 36-minute walk with no operating GPS, a 1-hour and 19 minute weight-training session is shown, as well as a 30-minute walk and a 5-minute treadmill run, which also shows zero miles run. Endomondo does not estimate distance run on a treadmill like the Apple Watch Workout app does when you specify Indoor Run.

More elaborate statistics are available to people who purchase a premium membership, including graphs showing trends over time.

Training Plan

Training plans are a premium feature of Endomondo, meaning they are not available on the free version of the product. However, if you would like to prepare for a race, going premium could well be worth the cost. A training plan is tailored to your current level of fitness and is dynamically adjusted as you go along. You can train for anything from your first 5-kilometer race to a marathon. A plan can help you if all you want to do is increase your speed with no particular race in mind. An audio coach encourages you along the way.

Commitments

To keep yourself motivated, you can make a commitment to a certain distance, duration, or calorie goal, such as the one shown in Figure 9-14.

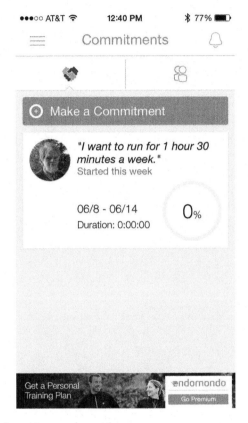

Figure 9-14. A commitment to a running goal

Endomondo will periodically remind you of how you are doing and display what percent of your goal you have already accomplished. You can also receive encouragement from your friends by sharing your commitment with them, along with your progress toward achieving the goal you have committed to. Your friends can also share their goals with you.

Challenges

Challenges are another way you can share mutual encouragement with your Endomondo friends. You can challenge specific friends to accomplish specific goals within a specific time frame at a specific sport. The competitive aspect makes it easier for you to motivate yourself to accomplish the goal of the challenge. Of course, if you have friends on Endomondo, they might challenge you to a goal of *their* choosing. In this way, everybody benefits. You can always decline to accept a challenge, if for some reason you do not want to engage in it.

Routes

Endomondo will store your favorite routes for you. You can create a route by tracing your route on www.endomondo.com and saving it. The app will calculate the distance and show it to you on a map, as shown in Figure 9-15.

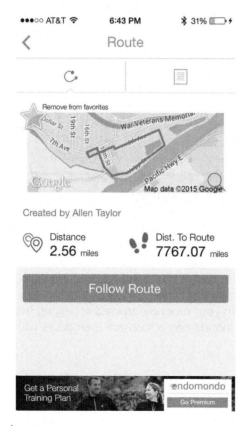

Figure 9-15. An example route

When you create a route, you can name it and specify whether it is private or public. If private, only you can see it. If public, any Endomondo member can see it and may challenge you to run that route in a faster time than they can run it.

One nice thing about public routes is the fact that if you are visiting some place and would like to take a run, you can be pretty sure that a public route listed on the Endomondo web site is a good one.

Settings

The Settings option on the Endomondo app is similar to the Settings option on other health and fitness apps. It enables you to set the parameters that you want your Apple Watch to display, including distance, heart rate, speed, duration, and average speed.

The Apple Watch Part of Endomondo

As is true for all the first-generation third-party Apple Watch apps, if the watch is separated from its paired iPhone by more than about 30 meters, it ceases to function. With the advent of watchOS 2, apps that are compatible with it will continue to work, as long as you are within range of a Wi-Fi signal. This is not much help when you are out on a trail run.

You can select the statistics that you want Endomondo to display during and after an exercise. Figure 9-16 shows the watch display after the five-minute treadmill run mentioned in the "History" section. Duration shows 5 minutes and 25 seconds. Distance on the treadmill shows 0.00 miles, and the app estimates that I burned 60 calories in the process.

Figure 9-16. Duration, Distance, and Calories display

Sweeping a finger across the screen from right to left shows additional statistics, although in this case they are not very helpful, as shown in Figure 9-17.

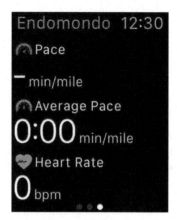

Figure 9-17. Pace, Average Pace, and Heart Rate screen

Because no distance was measured, the Pace metric shows a blank, and Average Pace shows zero minutes per mile. Heart Rate shows zero beats per minute because heart-rate monitoring was not being done. It's kind of a hassle to put on a heart-rate monitor chest strap, and after you do, it is not the most comfortable thing in the world. The fact that the Apple Watch measures heart rate without you even having to think about it is a real advantage. It is unfortunate that the first-generation third-party apps such as Endomondo cannot use the data from the watch heart-rate monitor.

RunKeeper

Although running is implied by the name RunKeeper, as is the case with Endomondo, RunKeeper will adjust the calorie burn rate it calculates, based on the activity you are doing. The choices are as follows:

Running

Cycling

Mountain Biking

Walking

Hiking

Downhill Skiing

Cross Country Skiing

Snowboarding

Swimming

Wheelchair

Rowing

Nordic Walking

Other

The iPhone Part of RunKeeper

As Figure 9-18 shows, the main RunKeeper screen on the phone shows where you are on the map and prompts you to start a run.

Figure 9-18. RunKeeper main screen

Type of workout, route, and music playlist can be specified at the top of the screen, and a menu sits at the bottom, along with an advertisement for RunKeeper Elite, the paid version of the app.

The menu at the bottom has options of Me, My Plan, Start, Friends, and Settings. The Me screen, shown in Figure 9-19, shows how many miles you ran last month and how many you have run so far this month.

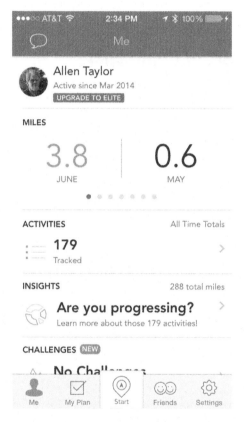

Figure 9-19. The Me screen

It shows how many activities you have recorded on RunKeeper and links to detailed records of all those activities. It also offers Insights as a premium feature, informs you of any challenges, charts workouts per week over the past couple of months, and lists your personal records.

The My Plan option takes you through a series of questions that determine what you want to achieve and where you are right now on the road to getting there. RunKeeper will then tailor a plan to your requirements, but you will need to be an Elite member in order to get it.

The Friends option informs you whenever a RunKeeper friend has performed an activity and also informs all your RunKeeper friends when you complete an activity. As you would expect, the Settings option enables you to set up RunKeeper the way you want it.

At the end of a run, RunKeeper shows you the route you have run along with your time, pace, and calories burned (Figure 9-20).

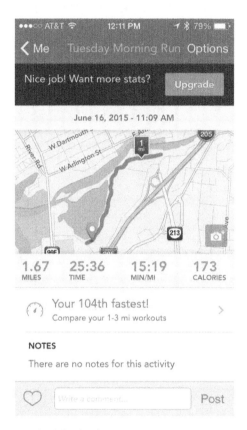

Figure 9-20. A short out-and-back by the river

In addition to time, pace, and calories, RunKeeper also gives you splits at mile marks and stacked graphs that show pace, elevation, and heart rate. Figures 9-21 and 9-22 show those three graphs.

Figure 9-21. Pace and elevation graphs

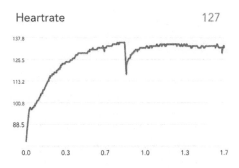

Figure 9-22. Elevation and heart rate graphs

The halfway point of the run was also the peak elevation. I stopped for a minute there, as chronicled by a drop in my heart rate. Resuming the run brought my heart rate back up to what it had been before the pause.

The Apple Watch Part of RunKeeper

The Apple Watch component of RunKeeper gives you all the information and control that you need before, during, and after the run. You can safely keep your iPhone in a pocket, without having to refer to it until later when you want to examine your stats. Figure 9-23 shows the Start screen.

Figure 9-23. RunKeeper Start screen

When you tap the Go Running button, a friendly voice emerges from your pocketed phone, announcing that the run has started. Along the way you can see how you are doing, as shown in Figure 9-24.

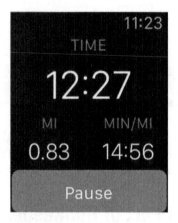

Figure 9-24. Screen while run is in progress

The Pause button enables you to pause the run. If you press it, you are given the choice of resuming or ending the workout. After you have ended it, you can either save or delete the workout. Later you can consult your phone at your leisure to look at saved workouts.

Strava

The Strava fitness app is specifically aimed at running and cycling, and you can switch from one to the other on the fly during a single workout. The capabilities are similar to those of the other apps described in this chapter, but the presentation of the information is somewhat different.

The iPhone Part of Strava

On the iPhone part of the Strava app, Figure 9-25 shows the start screen for running.

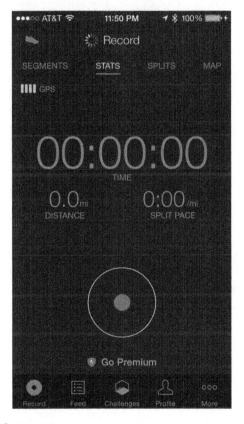

Figure 9-25. Running Start screen

As with the other apps in this category, the user is given the chance to "Go Premium" just below the circular button that you tap to start recording a run.

Figure 9-26 shows what you see at the end of a run.

Figure 9-26. Statistics at the end of a run

A map of the route taken is displayed, along with distance, time, average pace, and calories burned. Swiping from right to left shows the elevation profile of the run, as shown in Figure 9-27.

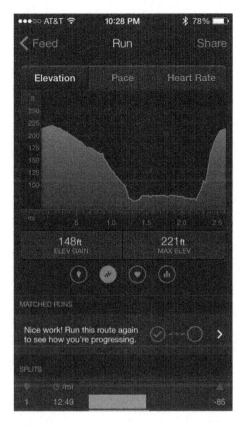

Figure 9-27. *Elevation profile*

The route displayed has a steep climb at the end of the course.

Tapping the Pace button shown in Figure 9-27 gives you a graph of your pace as it varied over the course. Figure 9-28 shows what this looks like.

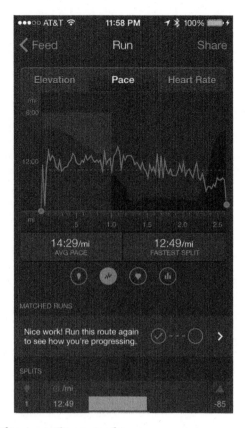

Figure 9-28. *Record of pace over the course of a run*

Split times appear in a bar graph at the bottom of the screen in Figure 9-29.

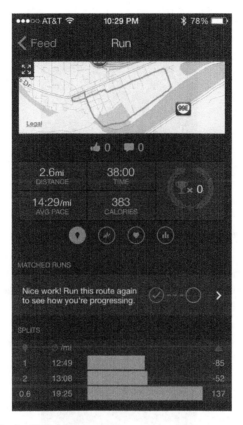

Figure 9-29. Bar graph of split times

As you would expect, the steep uphill stretch was a lot slower than the earlier part of the route.

The Apple Watch Part of Strava

The Apple Watch part of the Strava running and cycling app is rather minimalist. Figure 9-30 shows the start screen for running.

Figure 9-30. Start screen for running

The start screen for cycling is similar, except instead of a running shoe inside the circle, there is a bicycle. After a run is completed, the route taken is displayed on the watch, as shown in Figure 9-31.

Figure 9-31. Route displayed after a run

LifeSum

LifeSum is a health and fitness application, but it approaches those topics from a different perspective than the apps I have covered so far. Whereas the apps, both built-in and third party, that I have covered so far deal with time and effort spent exercising, LifeSum is primarily concerned with the flow of calories through a person's body. It records everything that a person eats and drinks in a day to arrive at a total number of calories consumed and balances that against calories burned through exercise. The beginning user of LifeSum enters their current weight and a target weight. LifeSum then generates a program that will take the person from their current weight to the target weight they want to reach. If users are below their calorie goal for the time of day, LifeSum encourages them to eat. If they have not drunk enough water, they are encouraged to down a glass of water. In addition to a daily calorie goal, consumption of carbohydrates, protein, and fats are also tracked and compared to target values. Ratios vary depending on the user's goals. A person wanting to lose weight would have different needs from a body builder.

The iPhone Part of LifeSum

After you have configured LifeSum on your iPhone, the Diary screen looks something like Figure 9-32, which is my screen at the beginning of a day before I have consumed anything.

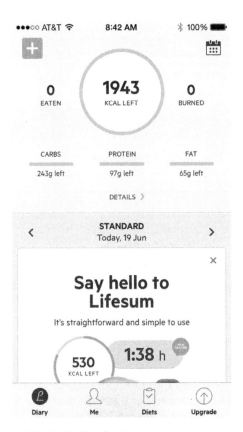

Figure 9-32. Diary screen at the beginning of a day

It shows the number of calories I should consume today to stay on track with my weight loss goal (1943) as well as the number of grams of carbs, protein, and fat that should be included in my day's diet. It also shows that so far I have consumed zero calories and have burned zero calories through exercise. The lower part of the scrollable screen shows the top part of the launch pad for a demo of LifeSum.

In addition to helping users track the calories they consume and burn, LifeSum also encourages them to keep adequately hydrated. Figure 9-33 shows eight empty 8.9-ounce glasses that will be filled as you press the plus sign to indicate that you have drunk the equivalent of a glass of water.

Figure 9-33. Water tracking section of Diary screen

Eight glasses is the minimum that LifeSum recommends that a person drink in a day. As you add records of exercise that you do, additional empty glasses will appear, indicating that your hydration needs have increased because of your increased activity.

LifeSum records what you eat and drink for breakfast, lunch, dinner, and snacks, as well as what exercise you do. It holds a large database of foods for which it knows the calorie content and a database of exercise activities along with the number of calories burned per unit time by engaging in those activities.

Note that if you want to eat a food that is not in the LifeSum food database, you can add it by using the LifeSum barcode scanner on your iPhone by pressing the SCAN A BARCODE button shown in Figure 9-34 and then aiming your phone at the barcode on the new food product.

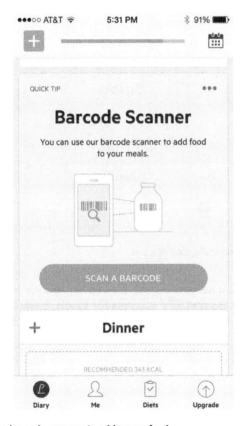

Figure 9-34. Using the barcode scanner to add a new food

Figure 9-35 shows the Diary section that covers lunch, dinner, snacks, and exercise before you have made any entries for the day.

Figure 9-35.   Diary section at the start of a day

Almost too faint to be seen in Figure 9-35 is the recommendation that I consume between 583 and 777 calories for lunch, between 758 and 991 calories for dinner, and between 0 and 389 snack calories.

You can enter what you have eaten in either LifeSum's desktop app or its mobile app. After you have entered breakfast, your Diary will look something like Figure 9-36.

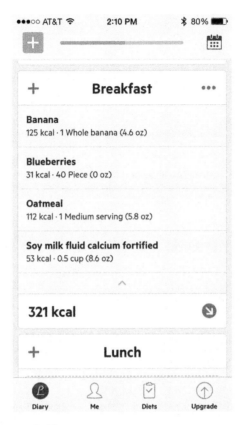

Figure 9-36. *Breakfast entry in Diary*

Each component of the meal is listed along with the number of calories it contains. Figure 9-37 shows the snack section of the Diary and shows that I have gone overboard on snacks today, packing away 699 calories that put me 48 calories over my target for this time of day (2:11 p.m.).

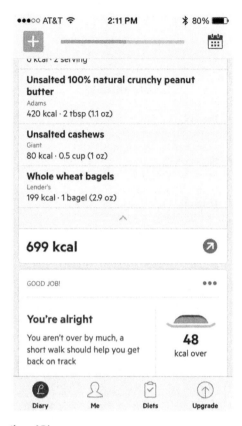

Figure 9-37. Snack section of Diary

If you have been faithful at making an entry in your LifeSum Diary every time you drink water as well as every time you eat anything, LifeSum will keep you up to date. Figure 9-38 shows that I had downed six glasses of water by 6:56 p.m.

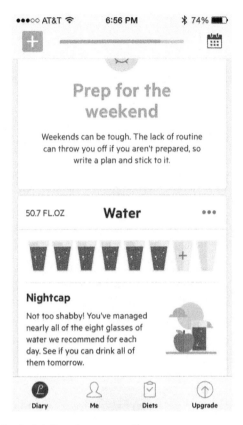

Figure 9-38. *Keeping track of daily water consumption*

Note   Don't take the recommendation of eight glasses of water every day as gospel. Different people have different hydration needs, based on their body type, how hot it is, how much they sweat, and how active they are. Also, any liquid such as tea, juice, or soda is almost entirely water and should count too. For sure, beer counts.

By the end of the day, as shown in Figure 9-39, I had added another snack, pushing my snack calorie total up to 869. I'm really going to have to do something about that. I had also done a little exercise, giving me 67 calories to balance against the 869 in snacks. I have to do better than that, and LifeSum will help me to do it by making the imbalance glaringly obvious.

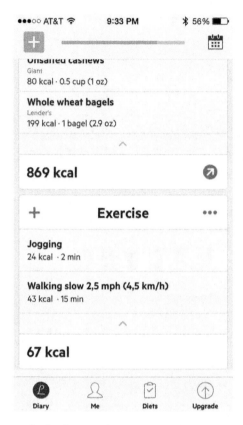

Figure 9-39. Too many snacks, too few exercises

At the end of the day, Figure 9-40 shows that I had consumed 1,647 calories but had burned only 67 in exercises. I have my eating pretty much under control but will have to do better in the exercise department.

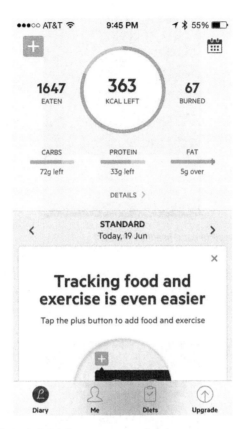

Figure 9-40. End-of-day calorie totals

The Apple Watch Part of LifeSum

The Apple Watch portion of the LifeSum app does two things. First, it enables you to enter your meals and exercise sessions without pulling out your phone. Second, it delivers timely reminders if you are under or over your calorie, water drinking, or exercise goals for a particular point in the day. Figure 9-41 shows a reminder screen that you would get early in the day.

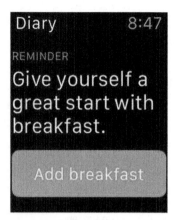

Figure 9-41. Encouragement to eat a healthy breakfast

After eating breakfast, when you tap the green "Add breakfast" button, the screen in Figure 9-42 appears.

Figure 9-42. How big was your breakfast?

You can specify that you have eaten either a small, a medium, or a large breakfast. Whether a breakfast is classified as small, medium, or large does not depend on the size of the breakfast but on its calorie content. Thus, a large bowl of oatmeal with skim milk would be a small breakfast, but a couple of pancakes with butter and maple syrup and a couple of sausages would be a large breakfast. Average calorie values are assigned to the small, medium, and large breakfasts, so the accuracy is not nearly as good as it would be if you entered the actual foods you ate into your iPhone, your laptop, or your desktop computer. If you want, you can enter the Small,

Medium, or Large description of your breakfast and edit it from your iPhone later to replace the value of the meal with the calorie counts of what you actually ate.

Once you have entered a value for a meal, your Diary will give you an "attaboy," such as the one in Figure 9-43, and tell you how many calories you have left to match your calorie consumption goal for the day.

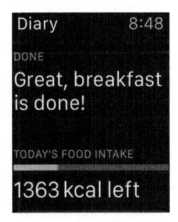

Figure 9-43. Breakfast data has been entered

Similar messages appear for other meals, snacks, and exercise sessions. I have found LifeSum to be helpful in keeping me within range of my food consumption goals, and it gives me subtle encouragement to shift my diet in a healthier direction. I can counter consuming too much fat by not putting quite so much peanut butter on my whole-wheat bagel. The bagel is actually pretty good without anything on it at all.

Runtastic

Runtastic is primarily a running app, similar to Nike+ Running, Endomondo, RunKeeper, and Strava. As you have seen, there are small differences between these apps, but the main functions are largely the same.

The iPhone Part of Runtastic

Like Nike+ and Endomondo, Runtastic has a main menu that you access by tapping the three stacked horizontal bars in the upper-left corner of your iPhone screen. Figure 9-44 shows most of the options. You can see the rest by dragging a finger up from the bottom of the screen.

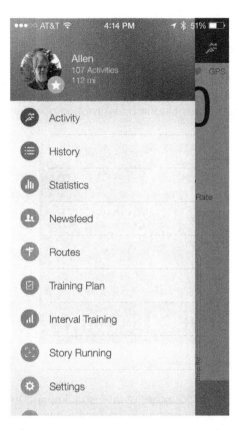

Figure 9-44. Runtastic main menu

Activity

The Activity option takes you to the screen you see at the start of a run or other activity (Figure 9-45). The pulsing blue dot on the map shows where GPS has located you. The Start button is at the bottom. When you press the Start button, your activity either starts immediately or starts after a delay that you are able to set, in seconds.

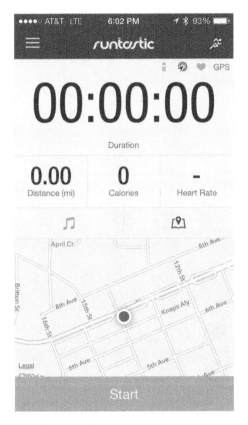

Figure 9-45. Activity screen at the start of a run

When the activity starts, a voice announces "Session started," and the clock starts counting seconds. There is now a slider at the bottom of the screen. When you drag it from left to right, the session is paused, and the voice helpfully says "Session paused." At this point, the slider is replaced by two buttons; the one on the left says Finish, and the one on the right says Resume. These do what you would expect. The Finish button ends the activity. The Resume button picks up where you left off when you dragged the Pause slider. Figure 9-46 shows the screen after tapping the Finish button.

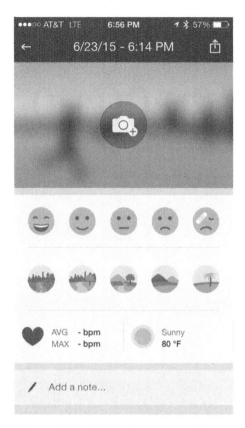

Figure 9-46. Characterize how you felt and what the course was like

By selecting an emoji, you can give a snapshot of how you felt and how challenging the route was. Tap the Done button when you are finished. At this point, the friendly voice will summarize the statistics of the run and display what has been recorded (Figure 9-47).

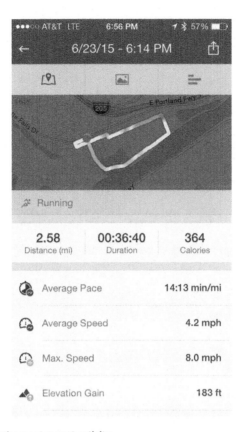

Figure 9-47. Stats for the most recent activity

Pressing the mountain icon in the top center will display weather information as well as the emojis that you have selected. Pressing the bar graph icon will display a graph of your speed throughout the course as well as its elevation profile. At the bottom, your splits are given along with elevation gain and loss during each of the splits. Figure 9-48 shows this data-packed screen.

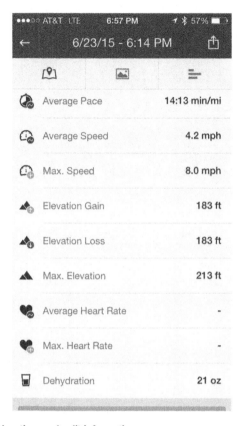

Figure 9-48. Speed, elevation, and split information

History

All the information recorded about an activity is retained in a History file. Figure 9-49 shows a recent History listing.

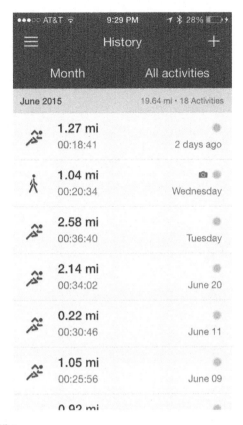

Figure 9-49. Recent History

By tapping whichever activity interests you, the same screen appears that you saw at the conclusion of that activity. The stats, weather, speed and elevation graph, and split table are also available.

Statistics

The Statistics screen (Figure 9-50) gives you your performance metrics for the previous month and a snapshot of how you have done so far this month.

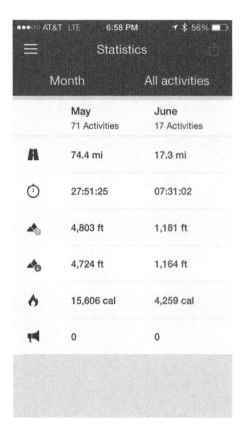

Figure 9-50. Statistics screen

It displays distance covered, time engaged in activities, elevation gain, elevation loss, kilocalories burned, and cheers received from your friends for your efforts.

Newsfeed

The Newsfeed option lets you share with your friends what you have been doing and enables you to see what they are doing. Figure 9-51 shows the time, weather condition, and route of a 1-mile walk. It also shows award icons for greatest elevation gain, fastest kilometer, and fastest mile walks.

Figure 9-51. *The Newsfeed option*

You can see what your friends who use Runtastic are doing. You can also make new friends by asking to see the activities of nearby users who may not be your friends already. Find someone who seems to be running the same distances that you are at about the same pace, and you may have found yourself a good running partner. Figure 9-52 shows what this looks like.

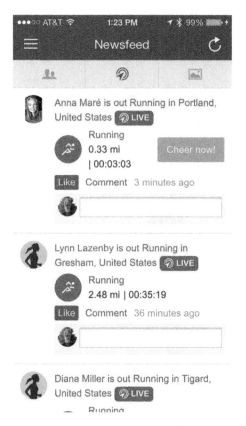

Figure 9-52. The activities of nearby runners who have been active recently

Another thing you can do is show your running friends (or walking friends, cycling friends, and so on) what the course looked like on your most recent activity. Figure 9-53 is an example, showing the road less traveled.

Figure 9-53. The road goes ever on

Routes

Runtastic users have the ability to add their favorite routes, whether for running, walking, cycling, or any other activity that involves going from one place to another, to a geographic database on the Runtastic web site. Just specify the route on a map by clicking points on the route. This is a great way to discover interesting routes that other people have found enjoyable or challenging. Figure 9-54 shows one of my favorite routes (in blue), along with pieces of routes that other people have entered (in black).

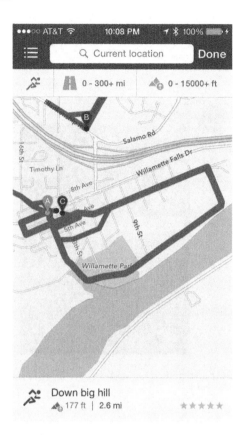

Figure 9-54. A pleasant 2.6-mile run

When you create a route, you can enter a description of it, and Runtastic will include a graph that shows the elevation profile of the route. Figure 9-55 is an example.

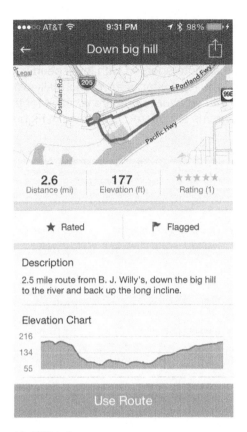

Figure 9-55. The "down big hill" route

Training Plan

Olympic champion Dieter Baumann has developed several training plans, including a weight loss plan that will help you to lose weight over a span of 22 weeks, and several running plans that range from plans for the complete beginner all the way to plans to prepare you to run (and finish) a marathon. Plans are aligned to how fast you want to run. In addition to the running training plans, there is a Bikini Body Prep plan for women, conducted by Lunden Souza, a certified personal trainer.

Training plans are free to Premium members of Runtastic.

Interval Training

Interval training is one of the best ways to increase speed and endurance. High-speed intervals are interspersed with low-speed intervals. During the high-speed intervals you run faster than you normally would, and during the low-speed intervals you recover to prepare yourself for the next high-speed interval. Several programs are offered, as shown in Figure 9-56.

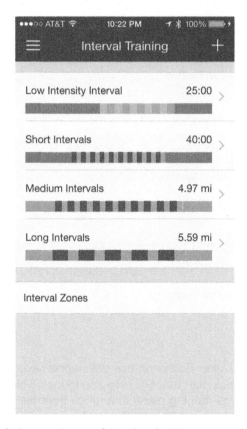

Figure 9-56. Progressively more strenuous interval workouts

Colors indicate speed, where blue is for easy warming up and cooling down, green is for slow running, yellow is for medium running, and red is for fast running.

Story Running

Story Running is a Premium feature. You can choose a story that runs somewhere between 30 and 40 minutes and listen to it through your headphones or earbuds while you run. The story could be science fiction, adventure, travel, inspiration, or something else, but the common denominator is that you are in the story and success depends on you to run. You could be running to escape ravenous wolves, to lose weight, or just to enjoy a run through the scenic beauty of Austria, described in detail by the program narrator.

Settings

The Settings screen enables you to customize the way the app behaves. Figure 9-57 shows you the types of things you can control.

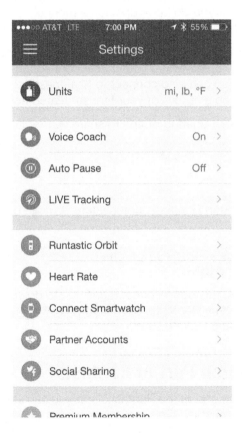

Figure 9-57. Settings screen

The Units option enables you to specify whether reports will be in terms of miles, pounds, and degrees Fahrenheit or in terms of kilometers, kilograms or stone, and degrees Celsius.

The Voice Coach option allows you to toggle the voice on or off that makes helpful announcements during a run.

GPS senses when you are stopped. If Auto Pause is enabled, every time you stop, Runtastic will pause the clock that is timing your run so that your stopped time does not lower the pace that is recorded. This saves you from manually pausing the clock when, for example, you stop to take a phone call.

LIVE Tracking lets you share an activity in real time with your Facebook friends or people on the Runtastic.com web site. It can send you cheers to keep you motivated while you are running or working out.

The Heart Rate option lets you hook up a chest-strap heart-rate monitor and sync it to Runtastic. If you do hook up such a monitor, it will send its data to the iPhone Health app just like your Apple Watch does.

Speaking of the iPhone Health app, the Partner Accounts option on the Settings menu is the place where you enable connecting Runtastic with the Apple Health app. You can also connect to MyFitnessPal here. Both are resources for retaining a record of your fitness-related activities.

You can use the Social Sharing option to share your activities on Facebook and Twitter.

Go to Shop

This option takes you to Runtastic.com, where you can buy more Runtastic merchandise.

More Runtastic Apps

As the name implies, this option encourages you to download other Runtastic apps, such as Runtastic Me, Road Bike, Mountain Bike, Sleep Better, Leg Trainer, and Butt Trainer.

The Apple Watch Part of Runtastic

With the Apple Watch part of the Runtastic app you can start, pause, and end a run, as well as elapsed time, distance covered, and heart rate if you have your chest-strap heart-rate monitor on and active. The first generation of Runtastic cannot access the heart-rate data generated by the Apple Watch. To see that, you must go to the Health app on your iPhone. Even if you control your activity entirely from your watch, the audible announcement of the start of the activity and of the statistics at the end will be made from your phone. Check to see whether your copy of Runtastic has been upgraded to work with watchOS 2 and, if so, what additional capability that brings.

Figure 9-58 shows the screen you see when you launch Runtastic.

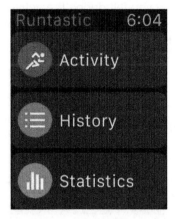

Figure 9-58. Main screen

The three things you can do are commence an activity, look at your past activities, and look at the statistics you have compiled since the beginning of the current month. When you press the Activity button, the screen shown in Figure 9-59 appears.

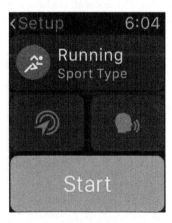

Figure 9-59. Activity screen

Press the green Start button and the activity will commence, either immediately or after a programmed delay that you have previously specified on your phone.

When you finish an activity, pressing the screen will activate Force Touch and display a screen giving you the option to either end or pause the activity. If you end it, the screen in Figure 9-60 will appear, giving you the opportunity to record how you felt during the activity and what kind of terrain your route took you over. You may also be given the option to share the activity on Facebook.

Figure 9-60. Specifying conditions during an activity

If you press the History button on the main screen shown in Figure 9-58, a display similar to Figure 9-61 will appear. It shows your three most recent activities and is scrollable so that earlier results can be displayed.

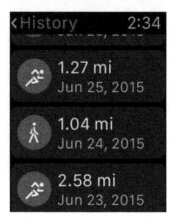

Figure 9-61. History display

Selecting Statistics from the main menu will show you something similar to Figure 9-62. It shows your totals for the current month in distance, duration, and elevation gain. Scrolling down will also show you elevation loss, calories burned, and cheers received from your friends.

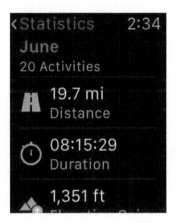

Figure 9-62. Statistics display

The main things you probably want to know about your current and previous activities are right there on your wrist. If you want more details, you can find them on your phone. Even more is available at Runtastic.com, which you can view on an iPad or PC.

Runtastic Six Pack

There are other kinds of fitness in addition to the aerobic fitness you get from running, walking, cycling, and other sports that get you breathing hard. One is to sculpt your body into the best possible shape. Perhaps the most iconic example is someone with "six-pack" abs.

Six-pack abdominal muscles are not easy to come by. The abdominal muscles must be strengthened and toned, but they also must be uncovered. Most people have a layer of fat that hides any six-pack that might be lurking underneath. Getting rid of excess fat is accomplished through diet, as covered by the LifeSum app earlier in this chapter. Strengthening and toning the abdominal muscles calls for physical movement exercises, which is what the Runtastic Six Pack is all about.

The iPhone Part of Runtastic Six Pack

The Runtastic Six Pack app goes beyond merely timing and recording an activity like other apps covered here do. It actually coaches you through a training program. The free version of the app introduces you to an avatar who performs the exercises that make up the training program. You are encouraged to perform the same exercises right along with the avatar. The main menu includes a number of helpful tools to move you along on your way to six-pack abs. Figure 9-63 shows the menu.

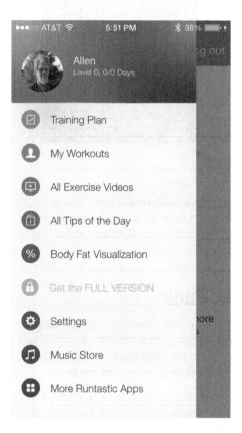

Figure 9-63. Main menu

Let's look at each of these options in detail.

Training Plan

Figure 9-64 shows the initial Training Plan screen. It describes the Level 1 training plan that will get you started with three sets of three exercises, with eight repetitions in each set.

Figure 9-64. *First Level 1 exercises*

The first exercise is the Tabletop Crunch + Twist (Figure 9-65). Your virtual trainer, either Daniel or Angie, performs the exercise, and you follow along. Your cadence is called by a narrator's voice, and appropriate music is played in the background.

Figure 9-65. A snapshot in the middle of the Tabletop Crunch + Twist

After three sets, you move along to the Hip Drops exercise (Figure 9-66).

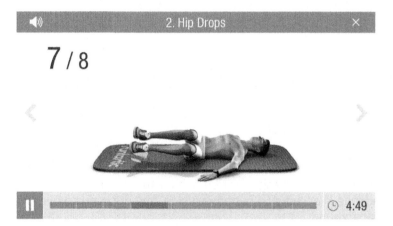

Figure 9-66. Hip Drops exercise when twisted to the left

The Crunch is the final exercise in this phase of Level 1. Figure 9-67 shows Daniel in the midst of performing it.

Figure 9-67. The Crunch

After completing these three exercises the prescribed number of times, you move on to new exercises. There are seven in all that comprise Level 1, which you will cycle through in ten days. To move on to Levels 2 and 3, you must purchase the full version. The seven Level 1 exercises will clearly be good for you, but 50 are available to purchasers of the full version, which can be had for a reasonable purchase price.

My Workouts

After you have gone through the training plan, to keep your momentum, you may decide to perform one or more of the predefined workouts described under the My Workouts option. Figure 9-68 shows the first of them and lists the names and durations of all five. All of these workouts are available only with the full version of Six Pack. The first, the Get in Shape workout, can be completed in less than nine minutes. The remaining four workouts take anywhere from 6 to 29 minutes to perform.

Figure 9-68. My Workouts iPhone screen

All Exercise Videos

There are three levels of exercise videos: Easy, Medium, and Hard. Seven of the easy ones are available to users of the free version, but the rest of the easy ones and all the medium and hard ones are available only to purchasers of the full package. Figure 9-69 shows the Easy exercises.

Figure 9-69. Easy exercise videos

All Tips of the Day

Helpful tips can make your journey smoother as you move toward your fitness goals, whether that includes building a six pack, running a new personal record for the mile, or lifting more weight than you ever have before. Currently there are more than 50 tips of the day that you can cycle through.

Body Fat Visualization

The Body Fat Visualization graph shows what your body should look like, based on your weight and percentage of body fat. The Runtastic Libra is a device that measures not only your weight but also your body fat and water percentages. Data is transferred via Bluetooth to your iPhone. Figure 9-70 shows the Daniel avatar as he would look with 13 percent body fat. Based on your actual measurements, made by the Libra, Daniel would be a stand-in for you, showing how you look with the most recent measurement of your percentage of body fat and weight.

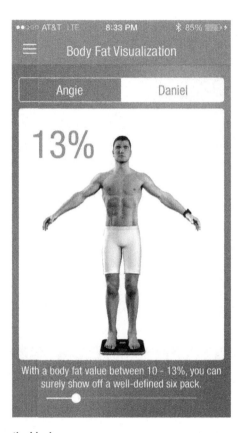

Figure 9-70. Visualizing the ideal you

Get the Full Version

This option is just a plug for upgrading to the full version of Runtastic Six Pack. Depending on your goals and how helpful you are finding the free version to be, you may decide to spend a few bucks and get the full package. It looks like it is worth the price to me.

Settings

Under the Settings option shown in Figure 9-71, you can select a training plan, a trainer, and a voice coach, and you can connect a chest-strap heart-rate monitor.

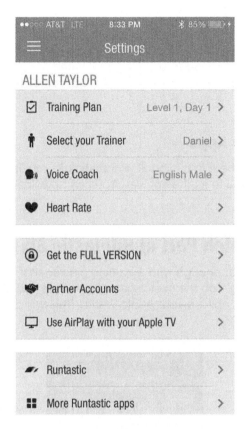

Figure 9-71. Settings menu

Naturally, Runtastic would like you to buy and connect its own brand of heart-rate monitor, but these devices all work the same way, so competing brands will work just as well.

Partner Accounts will let you connect with MyFitnessPal. If you do, Six Pack will share information with MyFitnessPal, and vice versa. MyFitnessPal can serve as a central repository for data about fitness that comes in from multiple apps, with Six Pack being only one of them.

The Use AirPlay with your Apple TV option will enable you to watch Daniel or Angie go through their paces on the big screen. This could be easier than trying to hold an iPhone while doing sit-ups.

Music Store

At the Runtastic Music Store you can buy music packs in a variety of genres that are all designed to keep you motivated and moving. One pack is provided for free so you can see what the packs are like without having to buy one first.

More Runtastic Apps

As I've noted elsewhere in this book, Runtastic has multiple apps, each aimed at helping you in a different way. You can download them from this menu option.

The Apple Watch Part of Runtastic Six Pack

Once you are familiar enough with a training plan so that you don't need to watch Daniel or Angie do the exercises while you are doing them, you can launch the Six Pack app on your Apple Watch. This will display the screen shown in Figure 9-72.

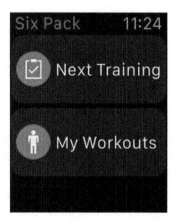

Figure 9-72. Six Pack main menu

Pressing the Next Training button will take you to the start of the training plan that you are currently working on. If you are at the beginning of Level 1, you will see the screen shown in Figure 9-73.

Figure 9-73. Level 1 training plan

When you tap the Start button, the voice coach on your phone will start talking you through the exercise in the same way as when you do the same exercise from your phone. The only difference is how you pause or end the session. Instead of tapping a control on your phone, press on the watch face to activate Force Touch. Options to pause and stop will appear, as well as the option to skip to the next exercise. If you elect to stop, when you come back later, Six Pack will return you to where you left off rather than taking you back to the beginning. Once you start on a training plan, your virtual trainer wants you to work through it to completion.

If you press the My Workouts option on the menu in Figure 9-72, a message will appear informing you that the workouts are available only in the full version of Six Pack. If you have the full version, then you can pick one of the five workouts described earlier in the iPhone section and shown in Figure 9-68.

Summary

For me, the real value of the Apple Watch as a tool to enhance health and fitness is delivered by the third-party apps rather than the built-in apps provided by Apple. This is even more true with the release of watchOS 2, which allows third-party apps to access the sensors on the watch. Even in their initial releases, the third-party apps provided a lot more information than the built-in apps do. The fact that this information is available on the fly while I am exercising, and can be accessed by a quick glance at an upraised wrist, is how the Apple Watch really earns its place on that wrist.

Chapter 10

What Could Be Improved?

No product is ever perfect. Whatever it is, it can be improved. This is especially true of the first release of a technical product such as the Apple Watch. Apple has worked hard to make the Apple Watch as useful as possible, while still being able to release the product while the market window was still open. There are some improvements that Apple will have to make to ensure the watch's continuing success. There are others that fall into the "nice-to-have" category. It will be interesting to see how many of these ideas ultimately are incorporated.

Increase Battery Life

Apple claims that with normal usage, the Apple Watch battery should last about 18 hours before it needs to be recharged. This means that some people will wear their watch from the time they wake up to the time they lay down at night with no problem. It also means that other people will continually run out of power before they run out of day. Apps use power, Bluetooth uses power, notifications use power, alerts use power, and flicking your wrist to illuminate the display uses power. The more things you use your watch for, the more power it will use. The bottom line is that the people who use their watches the most will run out of battery power the soonest and find themselves in the Power Reserve situation in which the functions they depend upon are unavailable and all they can do is read the time. Shortly thereafter, they can't even do that. Thus, the number-one thing that needs to be improved is the battery life. This is probably not something that can be accomplished with an application software update, although if an app is being a particular drain on the battery, it might be updated in a way that helps a little. It would be good to know, on an app-by-app basis, how much power each app is drawing. Perhaps Apple could provide that information, and users could decide whether they *really* need one that turns out to be a power hog.

Incorporate GPS into Apple Watch

Apple has marketed the Apple Watch as a health and fitness accessory, and indeed, fitness buffs have been early adopters of the device. People who want to track their activity, such as runners, walkers, or cyclists, are interested in knowing where they have been on a recent outing, rather than merely receiving an estimate of the number of steps they have taken. Location data requires GPS. The iPhone has GPS capability, and while you carry it, you can know the route you have taken. However, one of the great advantages of having a fitness watch for an athlete is that she does not have to carry her phone on a run or other workout. Unfortunately, when you leave your iPhone behind, you leave your GPS capability too. The Apple Watch will record how long you have been gone and estimate how many calories you have burned, but it will not tell you where you have been.

It's fine to say that Apple should add GPS capability to the Apple Watch, but there is a price to be paid for that capability. Adding GPS to the Apple Watch would add weight and take up space. This would likely require the Apple Watch to be bigger and bulkier. However, the biggest disadvantage is that the watch would draw more electrical power and run down the battery sooner. If it is true, as I claim, that the most needed improvement of the Apple Watch is to increase its battery life, adding GPS would be pushing things in the wrong direction. It's hard to see how Apple could both increase battery life and add GPS capability to the watch at the same time, without compromising the elegant design of the original product. If history is any indicator of things to come, Apple will not compromise the elegant design.

Connect Directly to the Internet, Regardless of Where You Are

Just as you must have your iPhone on your person to use GPS to track where you are, you also need it to access the Internet when you are out of range of a Wi-Fi node. This means that although it is convenient to use the Apple Watch to receive notifications, alerts, e-mails, and stock quotes, as well as many other functions, you will need to be within range of Wi-Fi to do so. Giving the watch all the major capabilities of an iPhone would be desirable, but it would entail added weight and bulk on the wrist and, most importantly, additional battery draw, causing the watch to run out of power sooner. Technology will have to advance before this capability is added to the watch.

Incorporate Blood Pressure Monitoring

Incorporating blood pressure monitoring is an additional capability that potentially will not require added weight or bulk because the hardware needed to perform this function is already included in the original version of the Apple Watch. However, the capability is not turned on. Apple has chosen to hold this capability back for reasons that I'm sure make sense to Apple. We will have to wait to see whether this capability is added at a later date through a software update or whether it will require a hardware modification. In the latter case, we are probably talking about buying a new watch, similar to the way we have needed to buy a new iPhone every time we want significantly upgraded capability.

Incorporate Oxygen Saturation Testing

The same hardware included in the Apple Watch that does blood pressure monitoring also does oxygen saturation testing. Because, in the initial release, this hardware is not activated, oxygen saturation is not monitored. If the capability is ultimately added, it will probably happen at the same time that blood pressure monitoring is added. As noted, this may be delayed until the technology improves, either by improving the accuracy of the reported results, by decreasing the amount of power this circuitry would draw if it were activated, or both. Oxygen saturation is a biometric value that has value to elite athletes who are trying to squeeze every possible ounce of performance out of their bodies. It would probably not mean much to casual exercisers who are just trying to maintain a healthy fitness level. This may be why Apple has not yet turned this feature on.

Activity App Algorithm Improvements

For the person who bought the Apple Watch primarily as an aid to health and fitness, the Activity app is the one that she will probably be referring to most often throughout the day. It is always running and always recording movement, exercise, and standing status. Thus, improvements in the Activity app will probably be the ones that watch wearers will notice the most.

Deducing Whether a Person Is Sitting or Standing

One of the truly useful features of the Apple Watch is its ability to sense that the wearer has been still for more than an hour and alert her that it is time to stand up for at least one minute. However, it is somewhat annoying to receive this alert when a person has already been standing continuously for the past hour. Apparently the sensor responsible for this function cannot differentiate between a person who has been sitting for an hour and a person who has been standing in the same place for an hour. You don't need to be a Buckingham Palace guard to stand unmoving for a whole hour. I often stand at my stand-up desk, doing my work, for more than an hour at a time. The Activity app tells me it is time to stand up anyway. I have found that I can satisfy it by leaving my desk and walking around in circles for a minute. This is probably just as well because I believe it is probably no healthier to stand in one place for an hour than it is to sit for the same length of time. Perhaps it would be better if rather than telling a person to stand for one minute, the Activity app would encourage her to move around for a minute.

Deducing Exercise to Include Nonaerobic Sports

Running, cycling, and even brisk walking clearly count as exercise. However, so do other things that do not necessarily transport your body from one place to another or raise your heart rate far above your normal level.

Vigorous weight lifting, where your muscles are fatigued to the extent that you cannot lift a weight even one more time than you already have, should count for something. However, such effort does not register on the Exercise ring of the Activity app. The Activity app should be improved to count more activities that should count as exercise but currently do not.

Add More Sports to Those Tracked by the Workout App

The Workout app tracks outdoor running, outdoor walking, outdoor cycling, indoor running, indoor walking, indoor cycling, stair stepper, rower, elliptical, and "other," which is assumed to burn calories at the same rate as brisk walking. These are many of the most popular aerobic exercises, but the functionality completely ignores anaerobic exercises, such as weight lifting. In addition, there is no provision for many recreational activities, such as soccer, football, baseball, bowling, golf, and basketball. Running is involved with a number of these but not on a continuing basis, so, for example, using the outdoor running option during a baseball game would give a calories burned estimate that is far from the number of calories actually burned. I'm not saying making an accurate estimate would be easy. Clearly, a pitcher is going to be burning a lot more calories than is a first baseman.

Summary

Clearly there are things about the Apple Watch that stand to be improved. Unfortunately, trade-offs are such that providing added capability in one area would require a decrease in capability in another. Just about any capability that you add will require additional power consumption, which implies a reduction in battery life, unless an entirely new type of battery is incorporated in the watch, one that has significantly increased power density. Such a "super battery" does not promise to arrive any time soon, so since extended battery life is probably the improvement that most people would want over any other, additional capabilities that require power are unlikely to appear in the near future. Having said that, additional capabilities that can be implemented by software upgrades may happen.

Setting Up Your Apple Watch

Because the Apple Watch is much more than just a watch, setting it up for the first time is somewhat more involved than just winding the crown as you would an old-time mechanical watch. The watch works hand in hand with your iPhone and must be paired with it so as not to be confused with any other nearby iPhones. Pairing your Apple Watch with your iPhone is covered in Chapter 3.

You will be asked to make some choices, such as whether you want to wear your watch on your left wrist or your right. I appreciate this option since I am one of the minority who prefer to wear their watch on their right wrist. The display is flipped 180 degrees from what it is for those who prefer their watch to be worn on the left wrist. With this capability, you can read it right side up regardless of which wrist the watch is on, and more importantly, you can easily operate the digital crown and the button next to it.

The My Watch Control Panel on the iPhone

Once you have activated your watch and paired it with your iPhone, you may want to customize a number of settings. The default settings will work well for most people most of the time, but you may find a few that you want to change. You can do this by tapping the Apple Watch icon on your iPhone's main screen. This will display the My Watch control panel shown in Figure A-1.

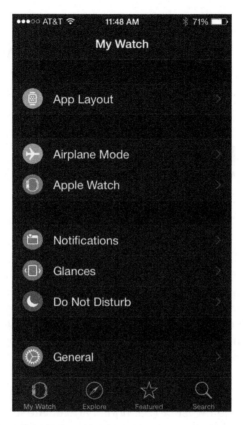

Figure A-1. My Watch control panel

Here you will see the categories of options that you can modify. Options are also available for the built-in apps, as well as the third-party apps that you may have. Third-party apps on your iPhone that have extensions for the Apple Watch will have been automatically updated to enable your watch to use them. It's a good idea to run through all these options at least once so that your watch is doing what you want it to do and displaying the information you want to see.

The Apple Watch Home Screen

The Apple Watch Home screen is the jumping-off point for whatever you want your watch to do for you. When you raise your wrist to look at your watch, it turns on and displays the time on the watch face that you have chosen from the many available. You can also customize the information that is shown on that watch face in addition to the time. Do any such customization with the My Watch app on your iPhone. Figure A-2 shows the default watch face.

Figure A-2. The default watch face

In addition to your local time, it displays the day of the month, your next calendar appointment, the temperature, an icon showing your activity level so far today, and the time in Cupertino, California, where Apple's headquarters is located. You can remove locations from the World Clock app (which is the source of this information) that you don't want or substitute locations that you *do* care about. Tapping the miniature Activity icon will take you directly to the Activity screen, home of the famous Activity rings. These rings will give you a quick picture of how active you have been so far today.

Index

W, X, Y, Z